Self-Organized Lightwave Networks

Self-Aligned Coupling Optical Waveguides

T0094393

Self-Organized Lightwave Networks

Self-Aligned Coupling Optical Waveguides

Tetsuzo Yoshimura

CRC Press
Taylor & Francis Group
Boca Raton London New York

CRC Press is an imprint of the
Taylor & Francis Group, an **informa** business

CRC Press
Taylor & Francis Group
6000 Broken Sound Parkway NW, Suite 300
Boca Raton, FL 33487-2742

© 2018 by Taylor & Francis Group, LLC
CRC Press is an imprint of Taylor & Francis Group, an Informa business

No claim to original U.S. Government works

Printed on acid-free paper

International Standard Book Number-13: 978-1-4987-7979-1 (Hardback)
978-1-138-74688-6 (Paperback)

**Visit the Taylor & Francis Web site at
http://www.taylorandfrancis.com**

**and the CRC Press Web site at
http://www.crcpress.com**

To the memory of my wife, Yoriko

also like to thank Drs. Hideyuki Nawata, Tetsuo Sato, and Juro Oshima of Nissan Chemical Industries, Ltd. for supporting the SOLNET research for a long time, and colleagues of Fujitsu Computer Packaging Technologies, Inc., San Jose, California, Fujitsu Laboratories, Ltd., and students of the Yoshimura Laboratory at Tokyo University of Technology for their collaboration in research work. Finally, the author would like to thank Ms. Ashley Gasque for giving me a great opportunity to write this book; Mr. Marc Gutierrez, editor; and Mr. Mario A. D'Agostino, editorial engineer of CRC Press/Taylor & Francis, and Mr. Paul Beaney, book manager of Nova Techset Private Limited for their instruction, support, editing, and typesetting in completing this book.

Author

Tetsuzo Yoshimura was born in Tokyo, Japan, in 1951, and graduated from Tokyo Metropolitan Aoyama High School in 1970. He received a BSc degree in Physics from Tohoku University in 1974, and MSc and PhD degrees in Physics from Kyoto University in 1976 and 1985, respectively.

In 1976, he joined Fujitsu Laboratories Ltd., where he was engaged in research on dye-sensitization, electrochromic thin films, amorphous super lattices, organic nonlinear optical materials, and polymer optical circuits. He invented the molecular layer deposition (MLD) and the self-organized lightwave network (SOLNET).

From 1997 to 2000, he worked at Fujitsu Computer Packaging Technologies, Inc. (FCPT), San Jose, California, where he planned the strategy on integrated optical interconnects within computers.

From 2001 to 2017, he was a professor at Tokyo University of Technology, where he extended the research on SOLNET-based optoelectronics and MLD-based nanotechnologies for optical interconnects, solar energy conversion systems, and cancer therapy.

He is currently a professor emeritus of Tokyo University of Technology, a scientific writer, and a musician in a rock group, The TYNC (Kekyo & Yoritan Records).

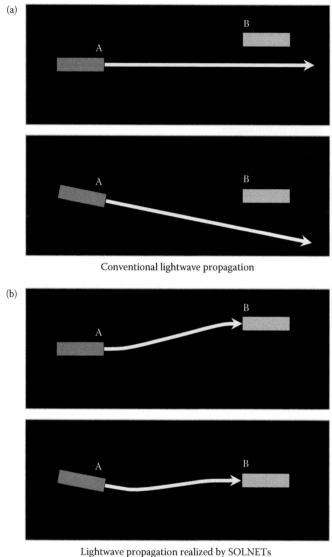

(a)

Conventional lightwave propagation

(b)

Lightwave propagation realized by SOLNETs

FIGURE 1.1   Lightwave propagation.

into the advanced OE systems, contributing to reductions of the system cost and energy dissipation.

In addition, SOLNETs will provide novel wiring processes by utilizing the capabilities to fabricate 3D optical wiring in free spaces and to target lightwaves onto specific objects, giving us a chance to create new OE architectures.

SOLNETs are expected to be applied not only to the advanced OE systems but also to any fields where lightwaves are utilized, such as solar energy conversion systems, cancer therapy, and so on.

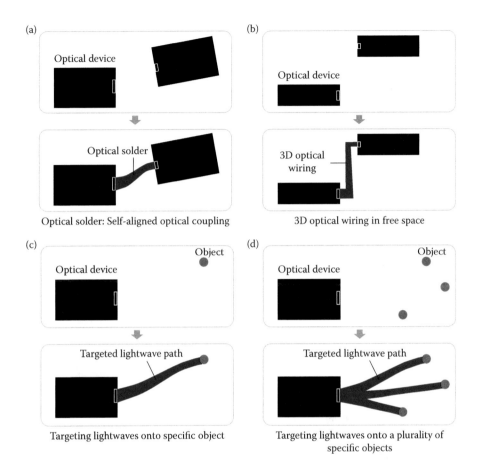

FIGURE 1.2 Capabilities of the SOLNET.

In this book, overviews of technologies related to the SOLNET, self-focusing of lightwaves in nonlinear optical media and self-written waveguides (SWWs), are given in Chapter 2. In Chapter 3, concepts and features are described for various types of SOLNETs. The performance of the SOLNETs is predicted by computer simulations based on the beam propagation method (BPM) and the finite-difference time-domain (FDTD) method in Chapter 4. After preferable waveguide growth condition for SOLNET formation is clarified in Chapter 5, experimental demonstrations are presented for various types of SOLNETs to show the proof of concepts in Chapter 6. In Chapter 7, expected applications of SOLNETs are proposed, including 3D integrated optical interconnects within computers, 3D micro optical switching systems, integrated solar energy conversion systems, and photo-assisted cancer therapy. Finally, future challenges are briefly discussed in Chapter 8.

2 Related Technologies

Prior to the main subjects of SOLNETs reported in Chapters 3–8, in this chapter overviews of related technologies, namely, self-focusing of lightwaves in nonlinear optical media and self-written waveguides (SWWs), are given.

2.1 SELF-FOCUSING OF LIGHTWAVES IN NONLINEAR OPTICAL MEDIA

Direct observation of the self-focusing of laser beams was reported by Garmire, Chiao, and Townes in 1966 [1]. When a Q-switched ruby laser beam of several tens of kW in power was introduced into a third-order nonlinear optical medium CS_2 through a 500-μm-diameter pinhole, the beam diameter decreased to 50 μm with the beam propagation, reaching a steady-state condition that generates a bright filament as schematically illustrated in Figure 2.1.

This result implies that the ruby laser beam is trapped in an optical waveguide formed by the beam itself. When the ruby laser beam intensity is I, due to the optical Kerr effect, the refractive index of the third-order nonlinear optical medium changes as

$$n = n_0 + n_2 I.$$

Here, n_0 and n_2 are respectively the linear refractive index and the optical Kerr constant. Because the intensity is high near the axis of the laser beam propagation as shown in Figure 2.1, the refractive index of the medium becomes high near the laser beam center compared to the surrounding region. This refractive index distribution confines the laser beam to be concentrated along the propagation axis, inducing self-focusing.

Similar self-focusing effects were observed in solids including crystals, glass, and optical fibers [2,3].

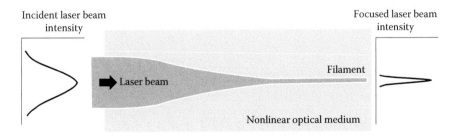

FIGURE 2.1   Schematic illustration of self-focusing.

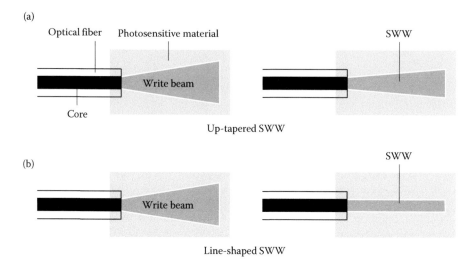

FIGURE 2.2 Schematic illustrations of SWWs.

2.2 SELF-WRITTEN WAVEGUIDES (SWWs)

The SWW was first reported by Frisken in 1993 [4]. A permanent optical waveguide was formed in a photosensitive material of a ultra-violet (UV)-cured epoxy by emitting a cw 532-nm laser beam from a core of a single mode optical fiber. Because the refractive index of the UV-cured epoxy increases by the laser beam exposure, the self-focusing effect similar to that in the third-order nonlinear optical media described in Section 2.1 was induced to produce a high-quality and low-loss up-tapered waveguide of SWW as schematically illustrated in Figure 2.2a. The optical waveguide stretching from the fiber core can be used as a low-cost beam expander which converts the spot size from ~10 to ~50 μm in diameter with a distance of a few millimeters. SWWs with straight-line shapes schematically shown in Figure 2.2b have been fabricated utilizing the self-focusing effect. They can be produced in photopolymers [5–9] and photosensitive glass [10].

Fazio et al. succeeded in SWW formation in photorefractive crystals to realize 3D optical interconnections in bulk ion-doped lithium niobate crystals by forming soliton waveguides [11–13]. The soliton waveguides construct light-induced integrated structures, which are permanently fixed, as well as digital and analog switching gates, opening a new approach toward novel integrated photonic circuits.

REFERENCES

1. E. Garmire, R. Y. Chiao, and C. H. Townes, "Dynamics and Characteristics of the Self-Trapping of Intense Light Beams," *Phys. Rev. Lett.* **16**, 347–349, 1966.
2. R. R. Alfano and S. L. Shapiro, "Observation of Self-Phase Modulation and Small-Scale Filaments in Crystals and Glasses," *Phys. Rev. Lett.* **24**, 592–594, 1970.
3. P. L. Baldeck, F. Raccah, and R. R. Alfano, "Observation of Self-Focusing in Optical Fibers with Picosecond Pulses," *Opt. Lett.* **12**, 588–589, 1987.

4. S. J. Frisken, "Light-Induced Optical Waveguide Uptapers," *Opt. Lett.* **18**, 1035–1037, 1993.
5. A. S. Kewitsch and A. Yariv, "Self-Focusing and Self-Trapping of Optical Beams upon Photopolymerization," *Opt. Lett.* **21**, 24–26, 1996.
6. M. Kagami, T. Yamashita, and H. Ito, "Light-Induced Self-Written Three-Dimensional Optical Waveguide," *Appl. Phys. Lett.* **79**, 1079–1081, 2001.
7. K. Dorkenoo, O. Crégut, L. Mager, and F. Gillot, "Quasi-Solitonic Behavior of Self-Written Waveguides Created by Photopolymerization," *Opt. Lett.* **27**, 1782–1784, 2002.
8. N. Hirose and O. Ibaragi, "Optical Component Coupling Using Self-Written Waveguides," *Proc. SPIE* **5355**, 206–214, 2004.
9. S. Jradi, O. Soppera, and D. J. Lougnot, "Fabrication of Polymer Waveguides between Two Optical Fibers Using Spatially Controlled Light-Induced Polymerization," *Appl. Opt.* **47**, 3987–3993, 2008.
10. T. M. Monro, C. M. de Sterke, and L. Poladian, "Investigation of Waveguide Growth in Photosensitive Germanosilicate Glass," *J. Opt. Soc. Am. B* **13**, 2824–2832, 1996.
11. E. Fazio, M. Alonzo, F. Devaux, A. Toncelli, N. Argiolas, M. Bazzan, C. Sada, and M. Chauvet, "Luminescence-Induced Photorefractive Spatial Solitons," *Appl. Phys. Lett.* **96**, 091107, 2010.
12. E. Fazio, A. Zaltron, A. Belardini, N. Argiolas, and C. Sada, "Influence of Iron Doping on Spatial Soliton Formation and Fixing in Lithium Niobate Crystals," *Opt. Mater.* **37**, 175–180, 2014.
13. E. Fazio, A. Belardini, L. Bastiani, M. Alonzo, M. Chauvet, N. I. Zheludev, and C. Soci, "Novel paradigm for integrated photonics circuits: transient interconnection network," *Proc. SPIE* **10130**, 1013006, 2017.

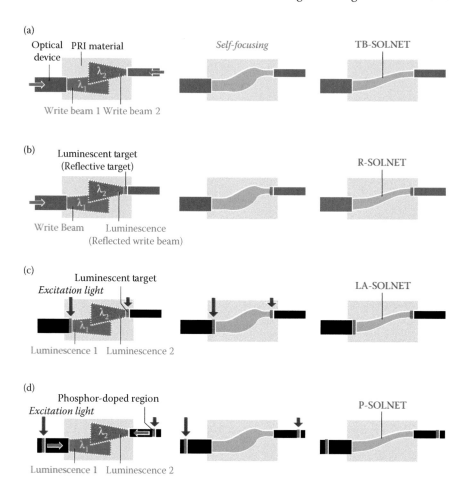

FIGURE 3.1 Types of SOLNETs. (a) Two-beam-writing SOLNET (TB-SOLNET), (b) Reflective SOLNET (R-SOLNET), (c) Luminescence-Assisted SOLNET (LA-SOLNET), and (d) Phosphor SOLNET (P-SOLNET).

luminescent target in the case that the waveguide core has emissive characteristics intrinsically. The R-SOLNET is effective when the write beam cannot pass through one of the optical devices.

The luminescent target in R-SOLNETs can be replaced with a reflective target such as a wavelength filter and a mirror to form R-SOLNETs by a write beam and a reflected write beam from the reflective target. In this case, λ_1 equals λ_2.

During the R-SOLNET formation, the luminescence from the luminescent target (or the reflected write beam from the reflective target) acts just like a trigger to let the incident write beam flow toward the emitting point (or the reflecting point). The luminescence (or the reflected write beam) paves a way to the target location to construct a self-organized waveguide toward the target. We call this phenomenon the "pulling water" effect.

In the LA-SOLNET [8], luminescent targets are placed on the frontcore edges of both optical devices. By introducing excitation lights from outside, say, from above, the targets respectively emit luminescence 1 of λ_1 and luminescence 2 of λ_2 to form a self-aligned coupling waveguide of LA-SOLNET. The sensitivity spectrum of the PRI material should be adjusted so that it exhibits low sensitivity to the excitation light and high sensitivity to the luminescence. The LA-SOLNET is effective when the write beams cannot pass through the optical devices.

In the P-SOLNET [1–4,8], phosphor-doped regions are formed in some parts of optical device cores to generate luminescence within the cores by excitation lights from outside. Luminescence 1 of λ_1 and luminescence 2 of λ_2 respectively correspond to the first write beam and the second write beam in the TB-SOLNET. They form a self-aligned coupling waveguide of P-SOLNET. The same situation is available without the phosphor-doped regions in the case that the waveguide cores have emissive characteristics intrinsically. In this case, the luminescence is generated just by exposing the waveguide cores to the excitation lights.

The four types of SOLNETs can be combined. When a P-SOLNET is combined with an R-SOLNET, namely, a luminescent target or reflective target is placed on the front core edge of one of the optical devices, a P/R-SOLNET is formed by luminescence generated in the other optical device. A P/TB-SOLNET is formed by a write beam from one of the optical devices and luminescence generated in the other optical device.

As schematically illustrated in Figure 3.2, in the R-SOLNET with luminescent targets, the luminescence from the luminescent target is radial while in the R-SOLNET with wavelength filters, the reflected write beam propagates to a direction deviating from the incident write beam. Consequently, the write-beam-overlapping effect is enhanced more remarkably in the R-SOLNET with luminescent targets than in the R-SOLNET with wavelength filters. In addition, as schematically illustrated in Figure 3.3, in the R-SOLNET with luminescent targets, by adjusting the sensitivity spectra of the PRI material, the reactivity of the PRI material at the wavelengths of the write beam and luminescence can be controlled. The adjustment of the sensitivity balance is equivalent to that of the intensity balance of the write beam and luminescence. An increase in the sensitivity at the wavelength of the luminescence corresponds to an increase in the intensity of the luminescence. Therefore, formation of the R-SOLNET with luminescent targets can be optimized by controlling the sensitivity spectra of the PRI material.

In TB-SOLNETs and R-SOLNETs, as illustrated in Figure 3.4a, the write beams should be introduced into the optical devices from outside using, for example, optical

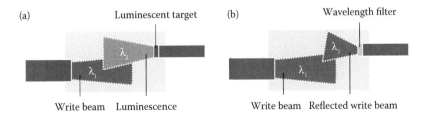

FIGURE 3.2 Beam overlapping in (a) R-SOLNET with luminescent targets and (b) R-SOLNET with wavelength filters.

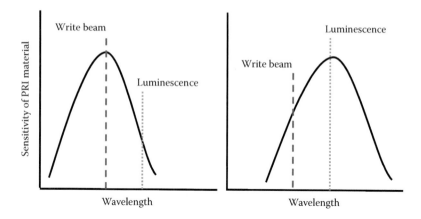

FIGURE 3.3 Adjustment of the sensitivity spectra of PRI materials in R-SOLNET with luminescent targets.

fibers with positional adjustments, which raise alignment costs, especially when the core size of the optical devices is nanoscale. In P-SOLNETs and LA-SOLNETs, on the other hand, the write beams are generated within optical devices by excitation lights from outside as schematically illustrated in Figure 3.4b. In this case, strict positional adjustments are not required, contributing to cost reductions for the SOLNET formation.

SOLNETs can provide branching structures. Figure 3.5a shows an example of Y-branching R-SOLNET formation. Luminescent targets are placed on the front edges of two optical devices. When a write beam from the opposing optical device enters the free space filled with a PRI material, the luminescent targets generate

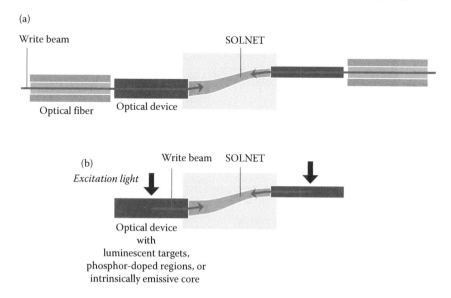

FIGURE 3.4 SOLNET formation apparatuses. (a) Write beams introduced from outside. (b) Write beams generated within optical device.

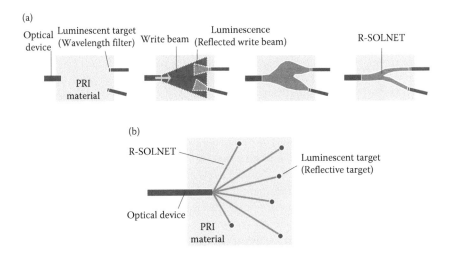

FIGURE 3.5   Examples of branching R-SOLNETs. (a) Y-branching R-SOLNET. (b) Multi-branching R-SOLNET.

luminescence to increase the refractive index in regions near them, guiding the incident write beam toward the target locations through the "pulling water" effect. Finally, by self-focusing, two self-aligned coupling waveguides connecting the left-hand-side optical device to the two right-hand-side optical devices are formed with a Y-branching structure. By using a write beam with a wide spread angle, a multi-branching R-SOLNET might be formed as shown in Figure 3.5b.

SOLNETs are expected to be formed by free-space write beams as schematically illustrated in Figure 3.6. In the example shown in Figure 3.6a, a free-space write beam

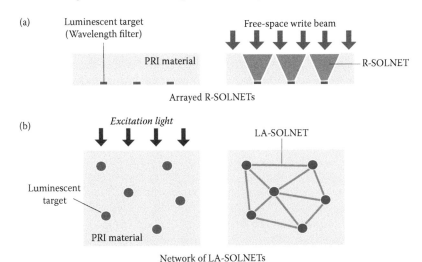

FIGURE 3.6   Examples of (a) tapered R-SOLNET array formed by a free-space write beam and (b) LA-SOLNET network formed by an excitation light.

is introduced into a PRI material, in which arrayed luminescent targets (or wavelength filters) are placed. The write beam is concentrated toward the arrayed targets to form arrayed R-SOLNETs having tapered structures. In the example shown in Figure 3.6b, by introducing excitation lights into a PRI material, in which a plurality of luminescent targets are placed, an LA-SOLNET network is constructed.

3.2 PRI MATERIALS

The requirement for PRI materials is that the refractive index increases upon write beam exposure. There are many kinds of PRI materials for SOLNET formation such as photopolymers [9–14], photodefinable materials, photosensitive glass [15], photosensitive organic/inorganic hybrid materials [16,17], photorefractive materials [18,19], and third-order nonlinear optical materials [20].

Figure 3.7a shows a typical mechanism of the refractive index increase in a photopolymer. High-refractive-index photoreactive molecules (high-n molecules) and low-refractive-index photoreactive molecules (low-n molecules) are mixed. The high-n molecules have higher photoreactivity to the write beam than the low-n molecules. When the photopolymer is exposed to the write beam, the high-n molecules are combined to make dimers, oligomers or polymers by photochemical reactions. Then, the high-n molecules diffuse into the exposed region from the surrounding region to compensate for the reduction of the high-n molecule concentration in the exposed region, and are combined to make dimers, oligomers, or polymers. This repeated process increases the refractive index of the exposed region to produce inhomogeneous refractive index distributions that are induced according to the pattern of the write-beam exposure. Curing by blanket light exposure or heating fixes the refractive index distributions to provide permanent SOLNETs.

The spectral response of the PRI materials can be adjusted by sensitizers. For the sensitizers, molecules with one-photon absorption characteristics or molecules with two-photon absorption characteristics can be used.

The gamma characteristics, that is, the relationship between the refractive index change and the write-beam exposure, can be controlled by adjusting the refractive index, photoreactivity, and concentration of molecules. Precise control of the gamma characteristics might be possible by mixing two, three, or more kinds of molecules with different refractive indices and photoreactivity as described in Chapter 8.

Figure 3.7b shows a refractive index increase mechanism in a photorefractive material such as lithium niobate (LN) crystals and photorefractive polymers, which exhibit the electro-optic (EO) effects. When the photorefractive material is exposed to a write beam, excited electrons are captured in electron traps, inducing local electric fields around the electrons. This gives rise to refractive index changes to produce inhomogeneous refractive index distributions that are induced according to the pattern of the write-beam exposure. Blanket exposure of erasing beams releases the trapped electrons to clear the refractive index distributions. This makes the material ready for rewriting, enabling dynamic SOLNETs.

Figure 3.7c shows a case in which a third-order nonlinear optical material such as polydiacetylene and compound semiconductors are used for the PRI material. When

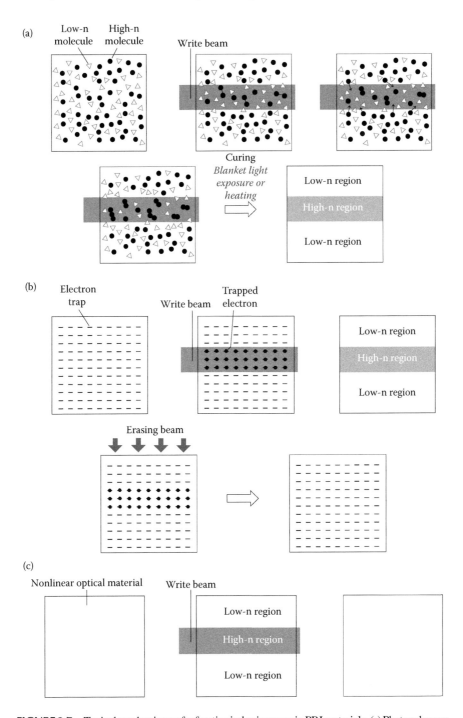

FIGURE 3.7 Typical mechanisms of refractive index increase in PRI materials. (a) Photopolymers for permanent SOLNETs, (b) Photorefractive materials for dynamic SOLNETs, and (c) Third-order nonlinear optical materials for dynamic SOLNETs.

the material is exposed to a write beam, inhomogeneous refractive index distributions are induced according to the pattern of the write-beam exposure due to the optical Kerr effect. The refractive index distributions disappear just after the write-beam exposure stops to reset the material. This enables dynamic SOLNETs.

3.3   ONE-PHOTON AND TWO-PHOTON SOLNETs

We call the SOLNET formed by conventional one-photon photochemistry as "one-photon SOLNET" and that formed by two-photon photochemistry as "two-photon SOLNET."

In the one-photon SOLNET, PRI materials exhibiting one-photon photochemistry are used. In this case, as shown in Figure 3.8a, the λ_1-write beam excites electrons from S_0 to S_n state in sensitizing molecules to induce chemical reactions, specifically, an increase in the refractive index of the PRI material. In parallel, the λ_2-write beam excites electrons from S_0 to $S_{n'}$ state to induce an increase in the refractive index. Therefore, the refractive index increase rate R is proportional to $\gamma_1 I_1 + \gamma_2 I_2$, and is expressed as:

$$R \propto \gamma_1 I_1 + \gamma_2 I_2, \tag{3.1}$$

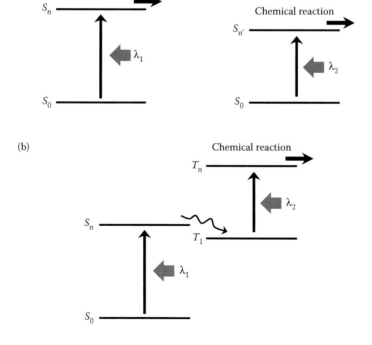

FIGURE 3.8 Energy-level schemes for sensitization with (a and b) one-photon and two-photon photochemistry.

where I_1 and I_2 are respectively the intensity of the λ_1-write beam and the λ_2-write beam, and γ_1 and γ_2 are respectively the PRI material sensitivity to the λ_1-write beam and the λ_2-write beam. In the case that $\gamma_1 \approx \gamma_2 = \gamma$, Equation 3.1 is simplified as:

$$R \propto \gamma(I_1 + I_2). \tag{3.2}$$

The one-photon SOLNET formation mechanism works in both cases where $\lambda_1 \neq \lambda_2$ and $\lambda_1 = \lambda_2$.

In the two-photon SOLNET [1,8], PRI materials exhibiting two-photon photochemistry are used. Two-photon photosensitive materials were reported by Brauchle et al. for holography [21]. Two-photon sensitizing molecules such as biacetyl (BA) and camphorquinone (CQ) were incorporated in the materials to induce two-photon photochemical reactions. The two-photon sensitization mechanism is shown in Figure 3.8b. Electrons excited from S_0 to S_n state by the λ_1-write-beam transfer to T_1 state, and are further excited to T_n state by the λ_2-write beam, resulting in a refractive index increase [21]. In this case, because two-step excitation occurs in series, the refractive index increase rate is proportional to $I_1 I_2$, and is expressed as:

$$R \propto I_1 I_2. \tag{3.3}$$

The above-described difference in the write-beam intensity dependence of the refractive index increase rate causes differences in formation characteristics between the one-photon SOLNET and the two-photon SOLNET. In Figure 3.9, SOLNET formation is schematically illustrated for cases with extremely large misalignments. When a SOLNET is attempted to be formed by the one-photon photochemistry, the refractive index increases with a rate proportional to $\gamma_1 I_1 + \gamma_2 I_2$. Because I_1 and I_2 are respectively large near the front edges of the left-hand-side and right-hand-side optical devices, the refractive index tends to increase rapidly in the PRI material near the front edges of the optical devices as shown in Figure 3.9a. This prevents the

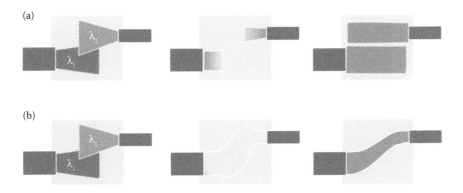

FIGURE 3.9 SOLNET formation schematically illustrated for cases with extremely large misalignments. (a) By one-photon photochemistry, (b) By two-photon photochemistry.

write beams from expanding into the write-beam-overlapping region in the middle, resulting in two separated optical waveguides stretching from individual optical devices. This means that the one-photon photochemistry cannot form SOLNETs when the misalignment is extremely large.

When a SOLNET is attempted to be formed by the two-photon photochemistry, on the other hand, the refractive index increases with a rate proportional to $I_1 I_2$, which means that the photochemical reactions occur only if the λ_1-write beam and the λ_2-write beam coexist. Then, as shown in Figure 3.9b, the increase in the refractive index near the front edges of the optical devices becomes less notable because I_2 is very small near the front edge of the left-hand-side optical device and I_1 is very small near the front edge of the right-hand-side optical device. In the write-beam-overlapping region in the middle, where both of I_1 and I_2 are not so small, the refractive index increase develops to make the two write beams merge to form a coupling optical waveguide of SOLNET. Thus, the two-photon SOLNET formation mechanism enhances the write-beam-overlapping effect to extend tolerances to misalignments.

The two-photon SOLNET also has an advantage that it can suppress the waveguide broadening caused by overexposure as illustrated in Figure 3.9 because the refractive index increase rate proportional to $I_1 I_2$ results in high-contrasting refractive-index images.

3.4 FABRICATION PROCESSES OF TARGETS AND PHOSPHOR-DOPED REGIONS

To form R-SOLNETs and LA-SOLNETs, luminescent targets should be deposited just on the core regions of optical device edges. The luminescent targets can be fabricated by the light curing process or the selective deposition process [22].

For optical devices consisting of optical waveguides, in which curing lights can be transmitted, the light curing process shown in Figure 3.10a is applicable. First, an edge of an optical device is coated with a luminescent material such as a phosphor-doped photopolymer. Then, in order to cure the region just on the core, the luminescent material is exposed to curing lights that propagate in the core of the optical device toward the optical device edge. The luminescent target is finally obtained by etching the noncured region. In principle, the target size is determined by the beam size of the propagated curing lights.

In the selective deposition process shown in Figure 3.10b, different surface properties are given to the core surface and the cladding surface of an optical device edge. For example, the core surface and the cladding surface are respectively hydrophobic and hydrophilic, which is the case of a Si waveguide with SiO_2 cladding. When a hydrophobic luminescent material is put on the surface, it is deposited on the hydrophobic core surface selectively to form a luminescent target. The luminescent material can be polymers, glass, or thin films.

The selective deposition process can also be achieved by surface treatments such as self-assembled monolayer (SAM) formation with designated patterns (see Appendix II).

Luminescent thin films can be grown by the vacuum deposition [7], the organic chemical vapor deposition [23,24], the atomic layer deposition (ALD) [25], or the

4 Performance of SOLNETs Predicted by Computer Simulations

In this chapter, the performance of SOLNETs is predicted by computer simulations. First, results of SOLNET formation and the coupling efficiency in optical couplings between microscale waveguides (micro–micro couplings) are presented. The simulations were performed by the two-dimensional beam propagation method (2D BPM), which is explained in Appendix I.

Next, results of SOLNET formation and the coupling efficiency in couplings between nanoscale waveguides (nano–nano couplings) and couplings between microscale and nanoscale waveguides (micro–nano couplings) with mode size converting functions are presented. The difference in the performance between the one-photon SOLNETs and the two-photon SOLNETs is clarified. The simulations were performed by the two-dimensional finite-difference time-domain (2D FDTD) method, which is explained in Appendix I. Since these types of SOLNETs involving nanoscale waveguides are especially important for applications to the optical solder in advanced OE systems, systematic discussions are attempted on them.

Then, effects of write-beam wavelengths, write-beam intensity, and gap distances between optical devices on SOLNET performance are described to discuss possibilities of widening misalignment tolerances in unmonitored SOLNET formation processes, where SOLNETs are formed at a fixed writing time.

In addition, SOLNETs with structural variations for specific applications such as optical Z-connections, high-density optical wiring, and cancer therapy are described, including vertical SOLNETs, parallel SOLNETs formed in arrayed structures, and Y-branching SOLNETs.

In all the simulations described in this chapter, the light absorption term in PRI materials is neglected. This implies that the simulations are for models under the small-absorption-limit condition, which is preferable for SOLNET formation as mentioned in Section 5.3.1.

4.1 SOLNETs BETWEEN MICROSCALE WAVEGUIDES

In the present section, results of simulations performed by the 2D BPM are presented for self-aligned optical couplings of TB-SOLNET, P-SOLNET, and R-SOLNET between microscale waveguides [1–5]. The results are only for one-photon SOLNETs. So, expressions of "one-photon" and "two-photon" are not marked.

4.1.1   TB-SOLNET/P-SOLNET

4.1.1.1   Couplings between Waveguides with Same Core Size

Simulation Models and Procedures

Figure 4.1a shows a simulation procedure based on the 2D BPM with paraxial approximation for the SWW, and Figure 4.1b shows that for the TB-SOLNET and

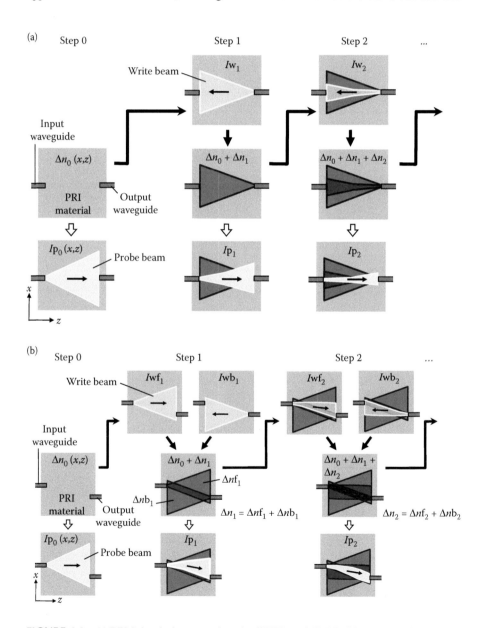

FIGURE 4.1   (a) BPM simulation procedure for SWWs and (b) TB-SOLNETs/P-SOLNETs.

the P-SOLNET. Because results for TB-SOLNETs and P-SOLNETs are the same as explained in Section 4.2.1, the results for TB-SOLNETs described below are those for P-SOLNETs at the same time.

In the simulations for SWWs and TB-SOLNETs, an input waveguide on the left and an output waveguide on the right are arranged in a counter-direction with a PRI material filling in between.

In the case of SWWs, as shown in Figure 4.1a, in step 0, an initial refractive index distribution is given as follows,

$$n(x,z) = n_0 + \Delta n_0(x,z),$$

where n_0 is a constant background refractive index. In the figure, only varying terms (Δn_i) are noted. A probe beam introduced from the input waveguide propagates in the PRI material with intensity of Ip_0 (x,z) under an influence of Δn_0. The probe beam does not affect the refractive index of the PRI material, and is used for assessments of optical couplings from the input waveguide to the output waveguide. In step 1, a write beam is introduced from the output waveguide into the PRI material. The write beam propagates with intensity of Iw_1 under Δn_0 to generate refractive index change Δn_1, resulting in a new distribution $\Delta n_0 + \Delta n_1$. A probe beam from the input waveguide propagates with intensity of Ip_1 under $\Delta n_0 + \Delta n_1$. In step 2, a write beam from the output waveguide propagates with intensity of Iw_2 under $\Delta n_0 + \Delta n_1$, resulting in a distribution $\Delta n_0 + \Delta n_1 + \Delta n_2$. A probe beam from the input waveguide propagates with intensity of Ip_2 under $\Delta n_0 + \Delta n_1 + \Delta n_2$. By repeating this procedure, SWW formation can be simulated.

In the simulation for TB-SOLNETs, as shown in Figure 4.1b, two write beams are considered. In step 0, a probe beam from the input waveguide propagates in the PRI material with intensity of Ip_0 (x,z) under Δn_0. In step 1, a write beam from the input waveguide (Iwf_1) and a write beam from the output waveguide (Iwb_1) propagate under Δn_0 to generate refractive index change Δnf_1 and Δnb_1, respectively, resulting in a new distribution $\Delta n_0 + \Delta n_1$. Here, $\Delta n_1 = \Delta nf_1 + \Delta nb_1$. A probe beam from the input waveguide propagates with intensity of Ip_1 under $\Delta n_0 + \Delta n_1$. In step 2, write beams from the input waveguide (Iwf_2) and from the output waveguide (Iwb_2) propagate under $\Delta n_0 + \Delta n_1$ to generate additional refractive index change $\Delta n_2 = \Delta nf_2 + \Delta nb_2$, resulting in a distribution $\Delta n_0 + \Delta n_1 + \Delta n_2$. A probe beam from the input waveguide propagates with intensity of Ip_2 under $\Delta n_0 + \Delta n_1 + \Delta n_2$. By repeating this procedure, TB-SOLNET formation can be simulated.

The refractive index of the PRI material is assumed to change according to the following relationship,

$$\Delta n = \alpha E^2, \tag{4.1}$$

where, E and α respectively denote the electric field of the write beam and a constant related to the PRI material sensitivity. The refractive index is assumed to increase until it reaches a saturation limit Δn_{sat}.

Parameters used in the simulations are summarized in Table 4.1. The width of the input and output waveguides is 8 μm. The gap distance between the waveguides

TABLE 4.1

Parameters in BPM Simulations for SOLNETs

Waveguide width	8 μm
Gap distance	400 μm
Background refractive index: n_0	1.5
Core-cladding refractive index difference	0.01
Write beam	Gaussian (480 nm, Half width: 3.6 μm)
	Peak intensity: 1.0 for SWW
	0.5 for SOLNET
Probe beam	Gaussian (1.3 μm, Half width: 3.6 μm)
	Peak intensity: 1

is 400 μm. The refractive index of the cladding region is 1.5, which is the same as n_0, and the core-cladding refractive index difference is 0.01, namely, the refractive index of the core is 1.51. The refractive index of the PRI material is 1.5 before write beam exposure. The saturation limit Δn_{sat} is 0.024. α in Equation 4.1 is 0.006 for calculations of SWWs and 0.003 for calculations of SOLNETs. Gaussian beams of 480 nm and 1.3 μm in wavelength are respectively used for the write beam and the probe beam.

SOLNET Formation

To compare SOLNETs to SWWs, first, we performed simulations of SWWs. Figure 4.2 shows SWW formation between the input and output waveguides with no lateral misalignments. The upper and lower contour diagrams respectively represent the refractive index and probe-beam intensity distributions. The two rectangles are the

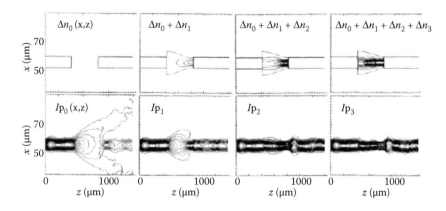

FIGURE 4.2   SWW formation between 8-μm-wide waveguides with no lateral misalignments. (From T. Yoshimura et al. "Self-Organizing Waveguide Coupling Method "SOLNET" and its Application to Film Optical Circuit Substrates," *Proc. 50th Electronic Components & Technology Conference (ECTC)*, 962–969, 2000.) [2]

input waveguide on the left and the output waveguide on the right. Note that the vertical axis scale is expanded.

A 3.6-μm-wide Gaussian write beam is introduced into the output waveguide to be propagated in the waveguide and emitted into the PRI material. A 3.6-μm-wide Gaussian probe beam is introduced into the input waveguide to propagate toward the output waveguide. It can be seen that, with write-beam exposure progress, a straight waveguide is formed gradually, which connects the input and output waveguides. As a result, the probe beam is guided from the input waveguide to the output waveguide.

When a lateral misalignment of 4 μm exists between the waveguides, as Figure 4.3a shows, considerable leakage of the probe beam remains even after the SWW is formed. Furthermore, the probe beam propagates with considerable meandering. When an angular misalignment of 2° exists between the waveguides, similarly, considerable probe-beam leakage occurs as Figure 4.3b shows, preventing efficient

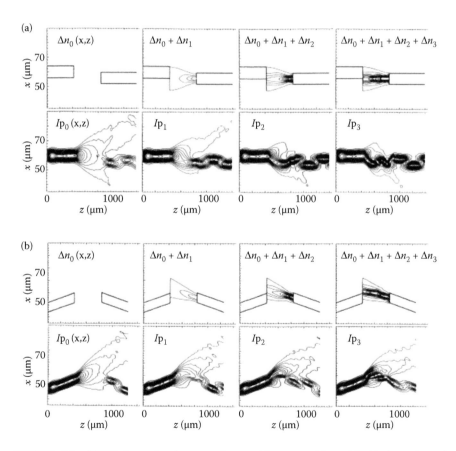

FIGURE 4.3 SWW formation between 8-μm-wide waveguides with (a) 4-μm lateral misalignment and (b) 2° angular misalignment. (From T. Yoshimura et al. "Self-Organizing Waveguide Coupling Method "SOLNET" and its Application to Film Optical Circuit Substrates," *Proc. 50th Electronic Components & Technology Conference (ECTC)*, 962–969, 2000.) [2]

probe-beam couplings between the waveguides. These results indicate that SWWs are not applicable to self-aligned optical couplings between misaligned waveguides.

Figure 4.4a shows TB-SOLNET formation between input and output waveguides with a lateral misalignment of 4 μm. Initially considerable probe-beam leakage is observed. With write-beam exposure progress, a self-aligned coupling waveguide of TB-SOLNET is gradually formed between the waveguides. Finally, the TB-SOLNET confines the probe beam to guide it from the input waveguide to the output waveguide efficiently. The probe-beam meandering, which is observed in the SWW, is suppressed in the TB-SOLNET to achieve a smooth probe-beam coupling with small mode disturbances.

Similarly, in the case that an angular misalignment of 2° exists between the waveguides, as shown in Figure 4.4b, a self-aligned coupling waveguide of TB-SOLNET is formed between the waveguides to couple the probe beam from the

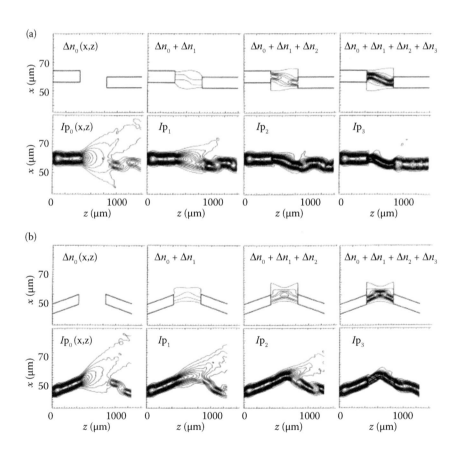

FIGURE 4.4 TB-SOLNET/P-SOLNET formation between 8-μm-wide waveguides with (a) 4-μm lateral misalignment and (b) 2° angular misalignment. (From T. Yoshimura et al. "Self-Organizing Waveguide Coupling Method "SOLNET" and its Application to Film Optical Circuit Substrates," *Proc. 50th Electronic Components & Technology Conference (ECTC)*, 962–969, 2000.) [2]

input waveguide to the output waveguide efficiently with suppressed probe-beam meandering.

The unique self-aligning characteristics of the TB-SOLNET arise from an interaction between the two write beams in PRI materials. The interaction is an attractive force, so that the two write beams merge together to produce one lightwave path.

Coupling Efficiency

In Figure 4.5a, coupling efficiency of TB-SOLNETs is plotted as a function of the exposure step count, that is, the writing time, for various misalignments. Initially, the efficiency increases with the exposure step count to reach a peak. After that, due to overexposure, the efficiency is reduced and reaches a steady state. With increasing the misalignments, the optimum writing time to reach the peak tends to increase.

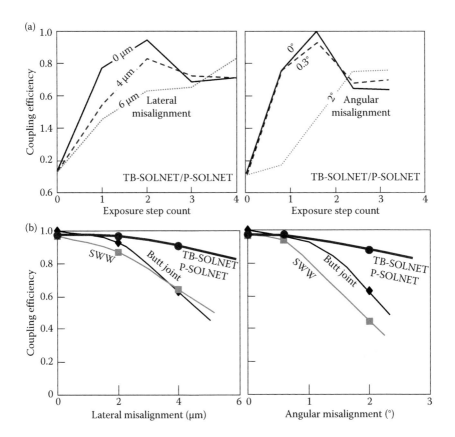

FIGURE 4.5   (a) Dependence of coupling efficiency on exposure step count in TB-SOLNETs/ P-SOLNETs formed between 8-μm-wide waveguides and (b) dependence of maximum coupling efficiency on lateral misalignments and angular misalignments for butt joints, SWWs, and TB-SOLNETs/P-SOLNETs. (From T. Yoshimura et al. "Self-Organizing Waveguide Coupling Method "SOLNET" and its Application to Film Optical Circuit Substrates," *Proc. 50th Electronic Components & Technology Conference (ECTC)*, 962–969, 2000.) [2]

Dependence of maximum coupling efficiency on lateral and angular misalignments is shown in Figure 4.5b for butt joints, SWWs, and TB-SOLNETs. In SWWs, the lateral misalignment tolerance is almost the same as that in butt joints, and the angular misalignment tolerance is narrower than that in butt joints. In TB-SOLNETs, on the other hand, both of lateral and angular misalignment tolerances are greatly widened comparing to those in butt joints and SWWs, keeping coupling efficiency above 80% (coupling loss: 1 dB) for a lateral misalignment of 6 μm, and an angular misalignment of 2°.

Thus, it is concluded from these results that the TB-SOLNET and P-SOLNET can form self-aligned coupling waveguides that connect misaligned microscale waveguides automatically, providing the optical solder function. This enables a probe-beam-leakage reduction as well as smooth propagation with small mode disturbances, resulting in coupling efficiency improvements.

4.1.1.2   Couplings between Waveguides with Different Core Sizes

TB-SOLNETs can be formed between microscale waveguides with different core sizes. Figure 4.6 shows TB-SOLNET formation between a 2.5-μm-wide input waveguide on the left and a 6-μm-wide output waveguide on the right with 4-μm lateral misalignment. The gap distance is 300 μm. A 1.4-μm-wide Gaussian write beam with peak intensity of 3.5 and a 3.3-μm-wide Gaussian write beam with peak intensity of 1.5 are respectively introduced into the input and output waveguides to be propagated in the waveguides and emitted into the PRI material. It is found that the write-beam exposure produces an efficient coupling waveguide of TB-SOLNET between the two waveguides, indicating that the TB-SOLNET and P-SOLNET are

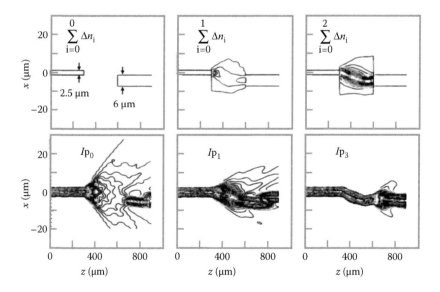

FIGURE 4.6   TB-SOLNET/P-SOLNET formation between 2.5-μm-wide and 6-μm-wide waveguides with 4-μm lateral misalignment. (From T. Yoshimura et al. *IEEE J. Select. Topics in Quantum Electron.* **9**, 492–511, 2003.) [4]

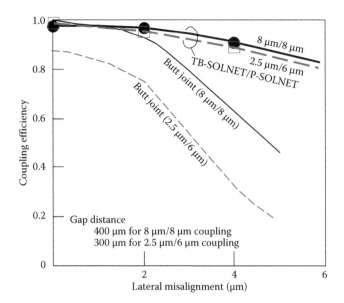

FIGURE 4.7 Dependence of maximum coupling efficiency on lateral misalignments for butt joints and TB-SOLNETs/P-SOLNETs.

effective for the mode size conversion between microscale optical devices with different core structures.

Figure 4.7 shows dependence of maximum coupling efficiency on lateral misalignments. TB-SOLNETs and P-SOLNETs exhibit wider lateral misalignment tolerances comparing to the butt joints.

4.1.2   R-SOLNET

Simulation Models and Procedures

A model for 2D BPM simulations of R-SOLNETs with wavelength filters is shown in Figure 4.8. Two 8-μm-wide waveguides are arranged in a counter-direction with a PRI material filling in between. The left-hand-side and right-hand-side waveguides are respectively input and output waveguides. The lateral misalignment is 2 μm and the gap distance is 240 μm. The refractive indices of the core and cladding region are 1.469 and 1.466, respectively. At the front edge of the output waveguide, a wavelength filter is attached. The filter transmits probe beams with a transmittance of 1 and reflects write beams with a reflectivity of 1. 680-nm write beams and 1.3-μm probe beams are introduced from the input waveguide. The write and probe beams have fundamental mode profiles with a peak height of $E = 1$. The refractive index of the PRI material is 1.466 before exposure, and increases with exposure according to Equation 4.1 with α of 0.003 and a saturation limit Δn_{sat} of 0.02.

The simulation procedure is shown in Figure 4.9. In step 0, refractive index change of the PRI material, Δn_0, is 0, which means that the refractive index is uniformly 1.466. A probe beam is introduced from the input waveguide for calculations of the

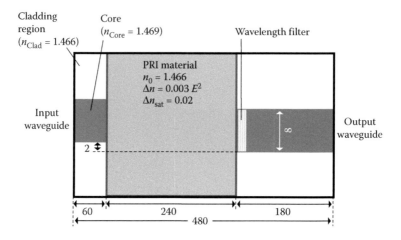

FIGURE 4.8   Model for BPM simulations of R-SOLNETs with wavelength filters between 8-μm-wide waveguides with 2-μm lateral misalignment (units: μm). (From T. Yoshimura et al. *J. Lightwave Technol.* **22**, 2091–2100, 2004.) [5]

electric fields, Ep_0, in order to evaluate the probe-beam coupling between the input and output waveguides.

In step 1, electric fields of a write beam, Ew_1, and electric fields of a reflected write beam from the wavelength filter, $Ew_{(R)1}$, are calculated using a configuration shown at the top in Figure 4.9. In the center of the PRI region, a light absorber is placed with a window corresponding to the wavelength filter aperture. The electric field distribution

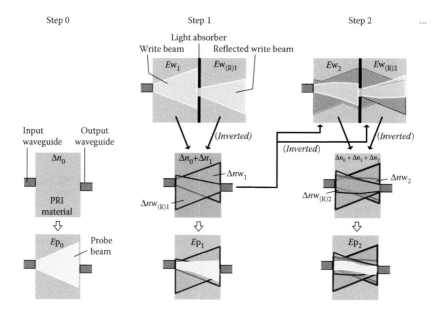

FIGURE 4.9   BPM simulation procedure for R-SOLNETs with wavelength filters. (From T. Yoshimura et al. *J. Lightwave Technol.* **22**, 2091–2100, 2004.) [5]

of the reflected write beam can be obtained by inverting the write beam distribution in the region on the right of the light absorber. Using Equation 4.1, the refractive index change distribution in the PRI material in step 1, Δn_1, can be calculated. Here, Δn_1 is a superposition of a refractive index change induced by the write beam, Δnw_1, and a refractive index change induced by the reflected write beam, $\Delta nw_{(R)1}$. Thus, total refractive index change becomes $\Delta n_0 + \Delta n_1$, which determines the electric fields of a probe beam, Ep_1. In step 2, Ew_2 and $Ew_{(R)2}$ are calculated, and then, refractive index change distribution in step 2, $\Delta n_2 = \Delta nw_2 + \Delta nw_{(R)2}$, is obtained, giving total refractive index change of $\Delta n_0 + \Delta n_1 + \Delta n_2$, which determines Ep_2. By repeating this procedure, simulations of R-SOLNETs with wavelength filters can be accomplished.

SOLNET Formation

In Figure 4.10, simulation results of the R-SOLNET formation are shown. Higher brightness represents larger refractive index or larger electric fields. In step 0, since refractive index distribution is uniform in the PRI region, a probe beam expands with propagation due to diffraction, inducing considerable leakage of the probe beam to the outside of the output waveguide. In step 1, a superposition of Ew_1 and $Ew_{(R)1}$ constructs a coupling waveguide, which reduces the probe-beam leakage. In step 2, self-focusing of the write beam and the reflected write beam becomes remarkable to construct a clear coupling waveguide between the input and the output waveguides, resulting in strong confinement of the probe beam.

For comparison, the same calculations were carried out for a model, where the wavelength filter is removed in the model shown in Figure 4.8. The model corresponds to a SWW model. The result is shown in Figure 4.11. Because no reflected write beam exists, only a straight waveguide stretching from the input waveguide is constructed. Calculated values of coupling efficiency from the input waveguide to the output waveguide for step 0, step 2 of the SWW, and step 2 of the R-SOLNET are respectively 58%, 76%, and 82%, indicating that the lateral misalignment tolerance in the R-SOLNET is wider than that in the SWW.

From these results, it is concluded that, although intensity of the reflected write beam is smaller than that of the incident write beam, the reflected write beam clearly contributes to pulling the coupling waveguide up to the exact position of the wavelength filter. The reflected write beam paves the way to the output waveguide to form the self-aligned coupling waveguide by the "pulling water" effect. In other words, the reflected write beam acts just like a trigger to let the incident write beam flow to the reflecting point. This enables the probe beam to reach the output waveguide.

Coupling Efficiency

In Figure 4.12a, coupling efficiency is plotted as a function of the exposure step count. Note that step 0 corresponds to the free-space coupling. In both of the R-SOLNETs with wavelength filters and SWWs, coupling efficiency increases with the exposure step count to reach a maximum, and decreases with further exposure. In the case of SWWs, the coupling efficiency reaches a steady state. In the case of R-SOLNETs, to reach the steady state, more exposure steps are necessary as can be seen in Figure 4.12b.

Figure 4.13 shows dependence of maximum coupling efficiency on lateral misalignments and gap distances for R-SOLNETs, SWWs, and free-space

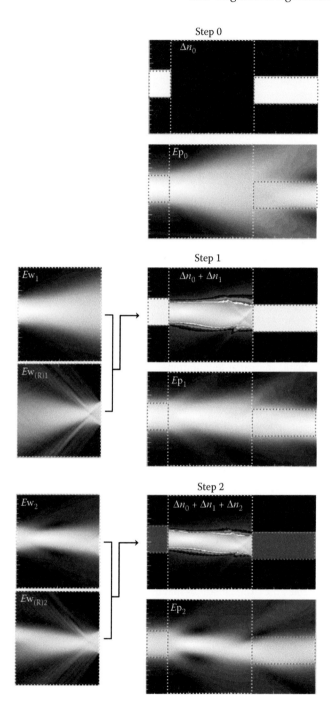

FIGURE 4.10 R-SOLNET formation between 8-μm-wide waveguides with a lateral misalignment of 2 μm. (From T. Yoshimura et al., *J. Lightwave Technol.* **22**, 2091–2100, 2004.) [5]

Step 2

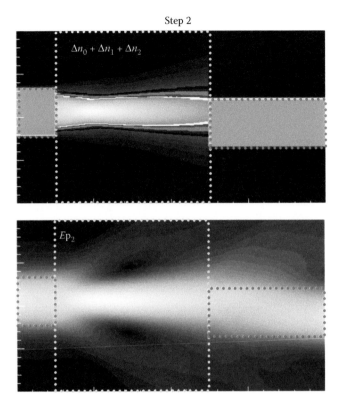

FIGURE 4.11   SWW formation between 8-μm-wide waveguides with a lateral misalignment of 2 μm. (From T. Yoshimura et al. *J. Lightwave Technol.* **22**, 2091–2100, 2004.) [5]

couplings. The R-SOLNET exhibits the highest coupling efficiency among the three. It achieves efficiency higher than 50% for 4-μm lateral misalignment, and exhibits a wider misalignment tolerance than the SWW. The usefulness of the R-SOLNET becomes remarkable particularly in the case of large gap distances. For 240-μm gap distance, the coupling efficiency at 2-μm lateral misalignment increases as ~35%, ~65%, and ~85% in the order of the free-space coupling, the SWW, and the R-SOLNET. With a decrease in the gap distance, the coupling efficiency in the R-SOLNET decreases due to the reduction in the overlapping between the incident write beam and the reflected write beam, resulting in small differences in coupling efficiency between the R-SOLNET and the free-space coupling in a small gap, for example, 60 μm.

By adjusting parameters such as the gamma characteristics of the PRI material, the coupling efficiency of the R-SOLNET with wavelength filters might be increased.

4.2   SOLNETs BETWEEN NANOSCALE WAVEGUIDES

In this section, results of simulations performed by the 2D FDTD method are presented for self-aligned optical couplings of TB-SOLNET, P-SOLNET, R-SOLNET,

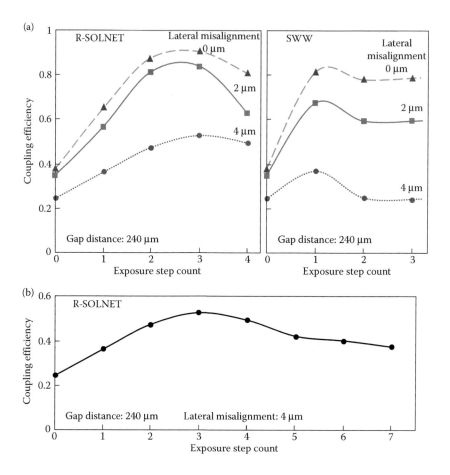

FIGURE 4.12 (a) Dependence of coupling efficiency on exposure step count in R-SOLNETs and SWWs. (b) Dependence of coupling efficiency on exposure step count in R-SOLNETs with 4-μm lateral misalignment for an extended range. (From T. Yoshimura et al. *IEEE Photon. Technol. Lett.* **17**, 1653–1655, 2005.) [13]

and LA-SOLNET between nanoscale waveguides [6]. The difference in the coupling performance between one-photon SOLNETs and two-photon SOLNETs is clarified.

4.2.1 SIMULATION MODELS AND PROCEDURES

Simulation Models

Models for 2D FDTD simulations of SOLNETs between nanoscale waveguides are shown in Figure 4.14. The width of the input and output waveguides is 600 nm. These optical waveguides are separated with a gap distance of 32 μm. The refractive index of the waveguide cores is 2.0 and that of the cladding region is 1.5. Such a refractive index distribution is available by using, for example, a sol-gel material [7] or SiON [8] for the cores and a polymer or glass for the cladding region. The refractive index of the PRI material increases from 1.5 to 1.7 upon write-beam exposure. Such a

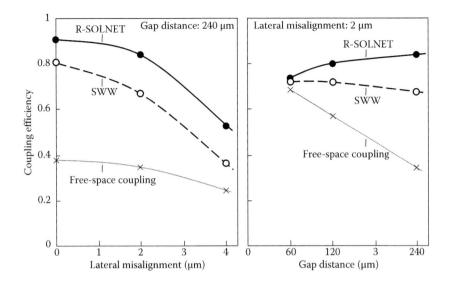

FIGURE 4.13 Lateral misalignment/gap distance dependence of coupling efficiency for R-SOLNETs, SWWs, and the free-space coupling. (From T. Yoshimura et al. *IEEE Photon. Technol. Lett.* **17**, 1653–1655, 2005.) [13]

photosensitive property is available in the PRI sol-gel material (Nissan Chemical Industries, Ltd.), which was found to be suitable for high-refractive-index-contrast SOLNET fabrication [9] as well as nanoscale waveguide fabrication [7]. The lateral misalignment between the input and output waveguides is expressed by d.

In the TB-SOLNET (Figure 4.14a), a λ_1-write beam is emitted from the input waveguide on the left and a λ_2-write beam from the output waveguide on the right. Thirty point light sources for the λ_1-write beam are distributed across the core along y-direction with a pitch of 20 nm at $x = 1.4$ µm in the input waveguide, and thirty point light sources

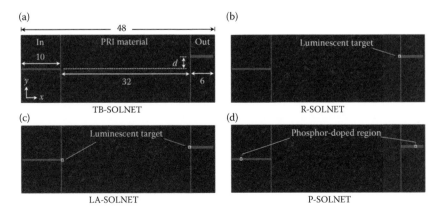

FIGURE 4.14 Models for 2D FDTD simulations of SOLNETs between 600-nm-wide nanoscale waveguides (units: µm). (a) TB-SOLNET (nano-nano), (b) R-SOLNET (nano-nano), (c) LA-SOLNET (nano-nano), and (d) P-SOLNET (nano-nano).

for the λ_2-write beam are distributed with a pitch of 20 nm at $x = 47.4$ μm in the output waveguide. Here, the origin of the x–y coordinate is set at the left-hand-side bottom corner of the model. The electric field of the point light source is expressed as follows,

$$\left. \begin{array}{ll} E(t) = \dfrac{1}{2}\left[1 - \cos\left(\dfrac{\pi t}{T_\omega}\right)\right]E_0 \sin \omega_0 t & 0 \leq t \leq T_\omega \\[4mm] E(t) = E_0 \sin \omega_0 t & T_\omega \leq t \end{array} \right\} \tag{4.2}$$

Here, ω_0 is the angular frequency of the lightwave and $\omega_0 T_\omega = 4\pi$. The electric-field amplitude of the write beam at the point light source location is $E_0 = 6$ mV/m.

In the R-SOLNET (Figure 4.14b), a 600-nm-wide line-shaped luminescent target is placed on the front edge of the output waveguide. When a λ_1-write beam is introduced into the PRI material from the input waveguide, the luminescent target is excited and emits luminescence of λ_2, which acts as the λ_2-write beam from the output waveguide. The point light source arrangements in the luminescence target are the same as those for the write beams in the TB-SOLNET. The electric-field amplitude of the emitted luminescence is determined by assuming that it is proportional to the temporary time average of absolute values of the λ_1-write-beam electric field at the location of the point light source [10,11]. Efficiency for the luminescence generation is assumed 0.7.

As mentioned in Section 3.1, the same situation for R-SOLNET formation is available without the luminescent target when the waveguide core has emissive characteristics intrinsically. In this case, the luminescent target in the model is regarded as the luminescent surface of the output waveguide core edge.

In the LA-SOLNET (Figure 4.14c), by excitation lights from outside, the λ_1-write beam and the λ_2-write beam are respectively generated from 600-nm-wide line-shaped luminescent targets on the front edges of the input and output waveguides. The point light source arrangements in the luminescent targets are the same as those for the write beams in the TB-SOLNET.

In the P-SOLNET (Figure 4.14d), by excitation lights from outside, the λ_1-write beam and the λ_2-write beam are respectively generated from a 600-nm-wide line-shaped phosphor-doped region at $x = 1.4$ μm in the input waveguide and from a 600-nm-wide line-shaped phosphor-doped region at $x = 47.4$ μm in the output waveguide. The point light source locations and the point light source arrangements in the phosphor-doped regions are the same as those for the write beams in the TB-SOLNET. In this case, the model for the P-SOLNET is exactly the same as that for the TB-SOLNET, indicating that results for P-SOLNETs and TB-SOLNETs are the same. In this book, therefore, the results for TB-SOLNETs imply the results for both TB-SOLNETs and P-SOLNETs.

As mentioned in Section 3.1, the same situation is available without the phosphor-doped regions when the waveguide core is emissive intrinsically. In this case, the phosphor-doped region in the model corresponds to the waveguide core portion, which is exposed to the excitation light.

To assess the coupling efficiency between the input and output waveguides, a probe beam of 650 nm in wavelength is propagated from the input waveguide to the output waveguide. The point light source locations and arrangements for the probe beam are the same as those for the λ_1-write beam.

Brauchle et al. reported that the absorption spectra of CQ, which is a two-photon sensitizing molecule, for λ_1 and λ_2 are respectively located in wavelength regions of 400–520 and 580–1100 nm (see Figure 6.28) [12]. According to these spectral data, we performed simulations for the following two cases; $[\lambda_1 = 500$ nm, $\lambda_2 = 600$ nm] and $[\lambda_1 = 400$ nm, $\lambda_2 = 780$ nm]. In this section, results for $[\lambda_1 = 500$ nm, $\lambda_2 = 600$ nm] are described. Results for $[\lambda_1 = 400$ nm, $\lambda_2 = 780$ nm] are described in Section 4.4 to discuss the influence of write-beam wavelengths on SOLNET formation.

Simulation Procedures

In the 2D FDTD calculations of SOLNET formation, after inputting the initial distribution of refractive index, the electric fields and the magnetic fields are calculated using the absorption boundary conditions. Because the refractive index of the PRI material changes with time, its distribution is updated at each time step n. Here, it is assumed that the photochemical reactions in the PRI material progress slowly, so that the delay time of the molecular diffusion can be neglected.

The energy density of the λ_1-write-beam exposure and that of the λ_2-write-beam exposure on the PRI material over a time interval for a step, Δt, are respectively $(1/2)$ $\varepsilon E_1{}^2 v\Delta t$ and $(1/2)\varepsilon E_2{}^2 v\Delta t$, where ε is the dielectric constant and v is the velocity of the lightwave. E_1 and E_2 denote the respective electric fields for the λ_1-write beam and the λ_2-write beam.

Then, for the one-photon SOLNET, since the rate of the refractive index increase is proportional to $\gamma_1 I_1 + \gamma_2 I_2$ according to Equation 3.1, the refractive index change Δn during Δt is expressed as,

$$\Delta n = \frac{1}{2}\varepsilon v\left(\gamma_1 E_1^2 + \gamma_2 E_2^2\right)\Delta t. \qquad (4.3)$$

In general, the PRI material sensitivity γ_1 and γ_2 are nonlinear functions of write-beam exposure, resulting in nonlinear gamma characteristics. In the present calculations, for simplicity, it is assumed that the refractive index changes in proportional to write-beam exposure until it reaches a saturation limit Δn_{sat} as schematically depicted in Figure 4.15, and γ_1 and γ_2 are set to the same value, γ. In this case, Equation 4.3 is simplified with a constant of proportionality:

$$\Delta n = C_{1Photon}\left(E_1^2 + E_2^2\right)\Delta t, \qquad (4.4)$$

$$C_{1Photon} = \frac{1}{2}\gamma\varepsilon v \qquad (4.5)$$

For the two-photon SOLNET, since the rate of the refractive index increase is proportional to $I_1 I_2$ according to Equation 3.3, Δn during Δt can be expressed as

$$\Delta n = C_{2Photon}\left(E_1^2 E_2^2\right)\Delta t. \qquad (4.6)$$

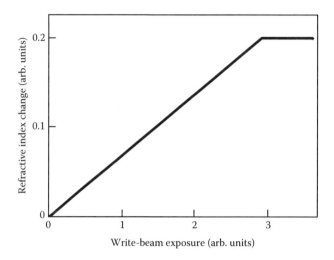

FIGURE 4.15 Schematic illustration of refractive index change vs. write-beam exposure characteristics in 2D FDTD calculations for SOLNETs.

The values of $C_{1Photon}$ and $C_{2Photon}$ are adjusted so that duration of SOLNET formation for two-photon SOLNETs is comparable to that for one-photon SOLNETs. This means adjusting the PRI material sensitivity. In the present study, we performed the simulations with constants of proportionality shown in Table 4.2.

The mesh sizes are $\Delta x = \Delta y = 20$ nm and $\Delta t = 0.0134$ fs, which were chosen to satisfy the convergence condition. The polarization direction is along z-axis that is perpendicular to the calculated plane. Because PRI materials are usually isotropic, polarization effects are likely to be negligible.

In real systems, the typical PRI material response time for SOLNET formation is over one second [1–5,13,14]. In the 2D FDTD method, however, it is difficult to perform calculations with such long time spans because Δt should be on a sub-fs scale. We therefore rescaled the time parameter in the results as follows: $\Delta t' = \Delta t \times 10^{15}$ and $\gamma' = \gamma \times 10^{-15}$. This corresponds to the FDTD calculations with $\Delta t = 0.0134$ fs referring to a PRI material with very high sensitivity.

4.2.2 BUTT JOINT

Figure 4.16 shows simulation results for butt joint couplings between 600-nm-wide waveguides. The left-hand-side and right-hand-side columns represent the dielectric

TABLE 4.2

Constants of Proportionality in 2D FDTD Calculations for SOLNETs

Simulation Model	$C_{1Photon}$ [$\times 10^{14}$ (m/V)^2/s]	$C_{2Photon}$ [$\times 10^{18}$ (m/V)^4/s]
TB-SOLNET/P-SOLNET	1.5	1.5
R-SOLNET	0.83	1.5
LA-SOLNET	2.6	9.1

n^2 E^2 (probe beam)

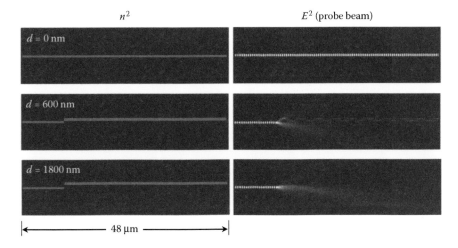

$d = 0$ nm

$d = 600$ nm

$d = 1800$ nm

|← —————— 48 μm —————— →|

FIGURE 4.16 Probe-beam propagation in butt joints between 600-nm-wide waveguides

constant ε, that is, the square of the refractive index n^2, and the intensity of the probe beams E^2 (probe beam), respectively. The probe beams with a wavelength of 650 nm are introduced from the input waveguide. Considerable leakage is observed for a lateral misalignment of 600 nm. For a lateral misalignment of 1800 nm, the probe-beam coupling becomes almost impossible.

4.2.3 TB-SOLNET/P-SOLNET

SOLNET Formation

As mentioned in Section 4.2.1, because the TB-SOLNET and the P-SOLNET are represented by the same model, the results described in this section for TB-SOLNETs are, at the same time, those for P-SOLNETs.

Figure 4.17a shows one-photon TB-SOLNET formation between 600-nm-wide waveguides with no lateral misalignments by write beams of $\lambda_1 = 500$ nm and $\lambda_2 = 600$ nm. The left-hand-side, middle, and right-hand-side columns respectively represent n^2, the intensity of the write beams E^2 (write beam), and E^2 (probe beam). A 650-nm probe beam is introduced from the input waveguide.

A straight waveguide is gradually formed with writing time t_W. The probe beam, which is initially diffuse, becomes concentrated into the output waveguide. At long writing time, the waveguide is broadened and wavy shapes appear due to overexposure as can be seen at writing time of 6.7 s. The waveguide broadening causes the probe beam to meander largely.

In the two-photon TB-SOLNET, as shown in Figure 4.17b, the broadening observed at 6.7 s is less notable compared with that in the one-photon SOLNET, and as a result, the probe-beam meandering is suppressed. This result is attributed to the rate of increase in the refractive index being proportional to $I_1 I_2$ in the two-photon SOLNET, which can form high-contrasting refractive-index images to suppress the broadening.

Figure 4.18a shows one-photon TB-SOLNET formation for a lateral misalignment of 3000 nm. Since the interaction between the write beams is not strong in the one-photon SOLNET, a coupling waveguide is unable to form. Instead, optical waveguides extend from the input and output waveguides with writing time to construct

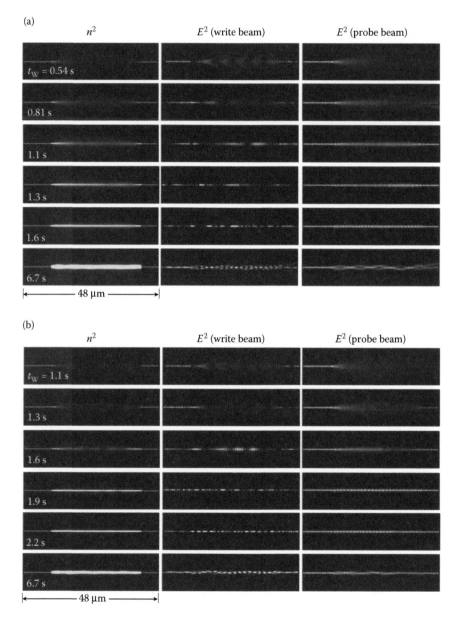

FIGURE 4.17 (a and b) One-photon and two-photon TB-SOLNET/P-SOLNET formation between 600-nm-wide waveguides with no lateral misalignments.

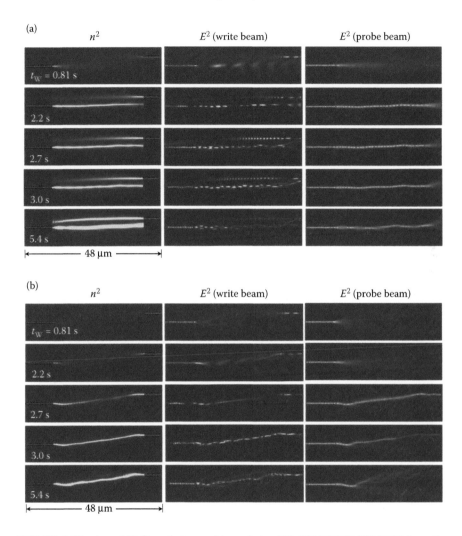

FIGURE 4.18 (a and b) One-photon and two-photon TB-SOLNET/P-SOLNET formation between 600-nm-wide waveguides with a lateral misalignment of 3000 nm.

completely-separated two optical waveguides. At the same time, considerable waveguide broadening is observed.

In the two-photon TB-SOLNET, on the other hand, as shown in Figure 4.18b, the two write beams begin to merge at 2.2 s to rapidly form a self-aligned coupling waveguide of two-photon TB-SOLNET. The probe beam is smoothly guided toward the output waveguide at 2.7 s. At 5.4 s, due to overexposure, zigzag shapes appear in the SOLNET, resulting in considerable probe-beam leakage.

Figure 4.19 shows n^2 profiles across TB-SOLNETs (along y-direction in Figure 4.14) at $x = 36$ μm, which corresponds to a position of 6 μm from the front edge of the output waveguide. The position $y = 16$ μm corresponds to the core axis location of the input waveguide. Graded-index-like profiles are observed at the initial stage

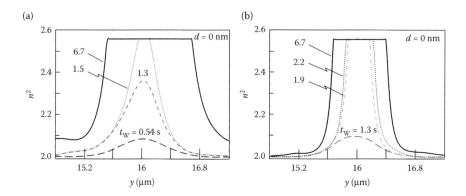

FIGURE 4.19 Cross-sectional n^2 profiles of (a) one-photon and (b) two-photon TB-SOLNETs/ P-SOLNETs at $x = 36\ \mu m$.

of SOLNET formation, and later change to step-index-like profiles. The waveguide broadening is suppressed in the two-photon SOLNET while considerable broadening occurs in the one-photon SOLNET.

The refractive index increase rate proportional to $I_1 I_2$ produces the steep two-photon SOLNET formation characteristics. As can be seen from Figure 4.19, one-photon TB-SOLNET formation begins at ~0.54 s and the time duration taken to reach its maximum n^2 value is more than 0.76 s while two-photon TB-SOLNET formation begins at ~1.3 s and the duration to reach its maximum n^2 value is much less than 0.6 s.

In Figure 4.20a one-photon TB-SOLNETs for various lateral misalignments are summarized. For a lateral misalignment of 600 nm, a self-aligned coupling waveguide is formed, and the probe beam is guided to the output waveguide. For lateral misalignments of 1800 and 3000 nm, normal SOLNETs cannot be formed. Instead, branching or parallel waveguides are formed, indicating that the tolerance to the lateral misalignment is ~600 nm.

In two-photon TB-SOLNETs, as shown in Figure 4.20b, the lateral misalignment tolerance is drastically widened up to 4200 nm, which is seven-times larger than the tolerance in the one-photon TB-SOLNETs. The waveguide broadening is suppressed, and consequently the meandering of probe beams in the output waveguides is suppressed, indicating that the two-photon SOLNETs induce small mode disturbances.

It should be noted that the two-photon TB-SOLNETs formed by $\lambda_1 = 500$ nm and $\lambda_2 = 600$ nm have symmetric S-shapes. The S-shaped structures give rise to wide lateral misalignment tolerances. Furthermore, the S-shaped structures suppress the probe-beam meandering to achieve optical couplings with small mode disturbances. The influence of the write-beam wavelengths on SOLNET formation is discussed in detail in Section 4.4.

Coupling Efficiency

Figure 4.21 shows dependence of coupling efficiency at 650 nm in wavelength on the writing time in TB-SOLNETs formed between 600-nm-wide waveguides. For

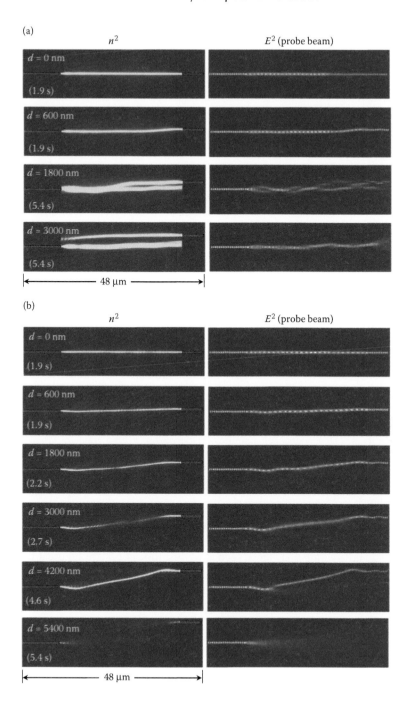

FIGURE 4.20 (a) One-photon and (b) two-photon TB-SOLNETs/P-SOLNETs between 600-nm-wide waveguides at optimum writing time indicated in brackets for various lateral misalignments.

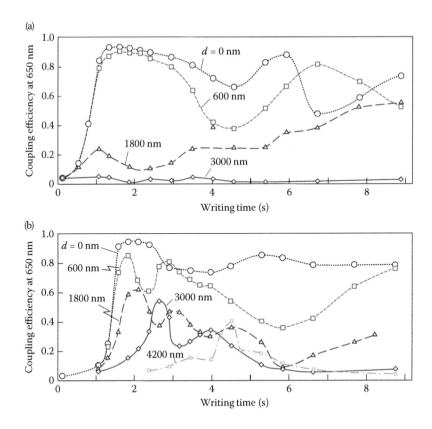

FIGURE 4.21 Dependence of coupling efficiency on writing time in TB-SOLNETs/ P-SOLNETs formed between 600-nm-wide waveguides. (a) One-photon TB-SOLNET/ P-SOLNET and (b) Two-photon TB-SOLNET/P-SOLNET.

one-photon SOLNETs shown in Figure 4.21a, when no lateral misalignments exist, the coupling efficiency increases to 94% with writing time, and then becomes unstable, varying between 47% and 88%. This instability of the coupling efficiency is due to the waveguide broadening that can be seen at writing time of 6.7 s in Figure 4.17a. For a lateral misalignment of 600 nm, the same trend is found in the coupling efficiency. For lateral misalignments of 1800 and 3000 nm, the coupling efficiency becomes very low, corresponding to the appearance of branching or separated waveguides with little sign of self-aligning (see Figure 4.20a).

For two-photon SOLNETs, as shown in Figure 4.21b, when no lateral misalignments exist, the coupling efficiency rises with writing time to 94%, and is maintained above 73% until 8.9 s. The instability of the coupling efficiency, which is observed in the one-photon SOLNETs, is suppressed. This is attributed to the suppression of the waveguide broadening as can be seen by a comparison of the one-photon and two-photon SOLNETs at 6.7 s in Figure 4.17. For a misalignment of 600 nm, the coupling efficiency reaches 84%, and thereafter tends to be unstable. For lateral misalignments of 1800 and 3000 nm, still relatively high coupling efficiency of 58%

and 53% is achieved, respectively. For a misalignment of 4200 nm, the maximum coupling efficiency is 40%, which is much higher than the coupling efficiency of less than 5% in the one-photon SOLNETs. These results confirm that the two-photon TB-SOLNET achieves greater lateral misalignment tolerances than the one-photon TB-SOLNET.

It is found from Figure 4.21b that with increasing lateral misalignments, the optimum writing time required to reach the maximum coupling efficiency tends to increase, and writing time windows keeping high-coupling efficiency becomes narrow. Such lateral misalignment dependence of the coupling efficiency prevents us from realizing unmonitored SOLNET formation. For practical use of the SOLNET optical solder, it is desirable that the optimum writing time is constant or the writing time windows are wide regardless of the lateral misalignments so that SOLNETs can be formed at fixed writing time without monitoring the coupling efficiency. Some approaches to solve this problem are discussed in Section 4.5 and Chapter 8.

In Figure 4.21, increases in the coupling efficiency are sometimes observed in the long writing time regions. This might be attributed to broadened, zigzagging, Y-branching, or directional coupling shapes of SOLNETs caused by overexposure. In such cases, a probe beam sometimes gets into the output waveguide with considerable efficiency accidentally. For example, the increase in the coupling efficiency observed for the one-photon TB-SOLNET with 1800-nm misalignment at writing time of 6–9 s is attributed not to the self-aligned coupling waveguide but to a directional coupler that incidentally appears in a particular parallel waveguide configuration (see Figure 4.20a).

Dependence of maximum coupling efficiency on lateral misalignments is shown in Figure 4.22 for TB-SOLNETs. Peak values of coupling efficiency in Figure 4.21 are plotted as a function of lateral misalignments. By introducing SOLNETs, the efficiency greatly increases from that in butt joints. In accordance with the results shown in Figure 4.20, the lateral misalignment tolerance is wider in two-photon

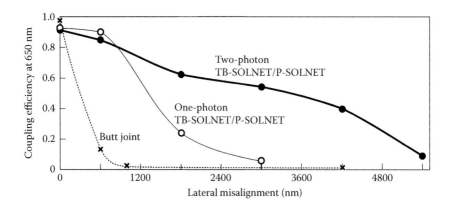

FIGURE 4.22 Dependence of maximum coupling efficiency on lateral misalignments in TB-SOLNETs/P-SOLNETs formed between 600-nm-wide waveguides.

TB-SOLNETs than in one-photon TB-SOLNETs. The wider misalignment tolerances of the two-photon SOLNETs are attributed to their capability to form high-contrasting refractive-index images as mentioned in Section 3.3 (see Figure 3.9). Because the rate of the refractive index increase is proportional to $I_1 I_2$, the refractive index increase occurs only if both of a λ_1-write beam and a λ_2-write beam coexist, resulting in the enhancement of the write-beam-overlapping effect to extend the misalignment tolerances.

4.2.4 R-SOLNET

SOLNET Formation

Two-photon R-SOLNET formation between 600-nm-wide waveguides with a lateral misalignment of 1800 nm is shown in Figure 4.23. When a λ_1-write beam is introduced from the input waveguide on the left, luminescence is emitted from a luminescent target, which is excited by the λ_1-write beam, on the front edge of the output waveguide. The write beam and the luminescence merge to form a self-aligned coupling waveguide at 4.0 s.

As Figure 4.24 shows, in one-photon R-SOLNETs, a self-aligned coupling waveguide is formed for a lateral misalignment of 600 nm. When the misalignment is increased to 1800 nm, a self-aligned coupling waveguide cannot be formed. In two-photon R-SOLNETs, on the other hand, a self-aligned coupling waveguide is formed even for a lateral misalignment of 3000 nm.

Coupling Efficiency

Dependence of coupling efficiency on the writing time (Figure 4.25) and dependence of maximum coupling efficiency on lateral misalignments (Figure 4.26) in R-SOLNETs are found to be almost similar to those in TB-SOLNETs. The lateral misalignment tolerance increases in the order of the butt joints, one-photon R-SOLNETs, and two-photon R-SOLNETs.

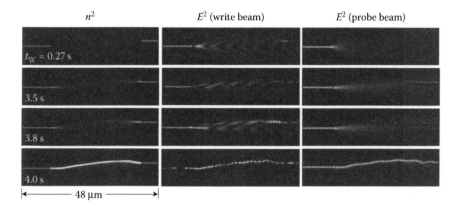

FIGURE 4.23 Two-photon R-SOLNET formation between 600-nm-wide waveguides with a lateral misalignment of 1800 nm.

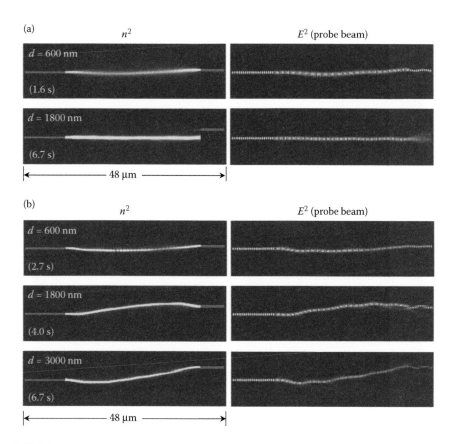

FIGURE 4.24   (a) One-photon and (b) two-photon R-SOLNETs between 600-nm-wide waveguides at optimum writing time indicated in brackets for various lateral misalignments.

4.2.5   LA-SOLNET

SOLNET Formation

Two-photon LA-SOLNET formation between 600-nm-wide waveguides with a lateral misalignment of 3000 nm is shown in Figure 4.27. Luminescence is emitted from both luminescent targets, which are deposited on the front edges of the input and output waveguides, to form a self-aligned coupling waveguide.

In one-photon LA-SOLNETs, as shown in Figure 4.28a, self-aligned coupling waveguides are formed for lateral misalignments of 600 and 1800 nm. For a lateral misalignment of 3000 nm, two separated waveguides are constructed instead of a coupling waveguide. In two-photon LA-SOLNETs, as shown in Figure 4.28b, a coupling waveguide is formed even for a lateral misalignment of 5400 nm, widening the lateral misalignment tolerance up to 5400 nm.

One concern in the two-photon LA-SOLNETs is that the coupling waveguide is not formed for a lateral misalignment of 4200 nm although it is formed for a lateral misalignment of 5400 nm. This indicates that, when the lateral misalignments

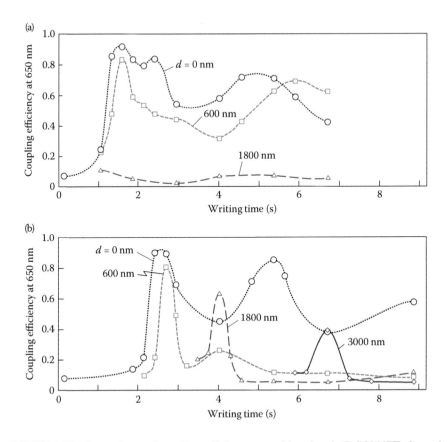

FIGURE 4.25 Dependence of coupling efficiency on writing time in R-SOLNETs formed between 600-nm-wide waveguides. (a) One-photon R-SOLNET and (b) Two-photon R-SOLNET.

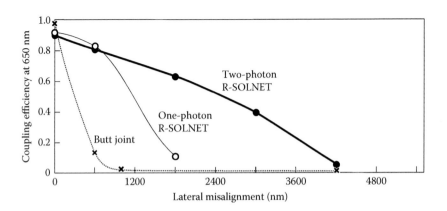

FIGURE 4.26 Dependence of maximum coupling efficiency on lateral misalignments in R-SOLNETs formed between 600-nm-wide waveguides.

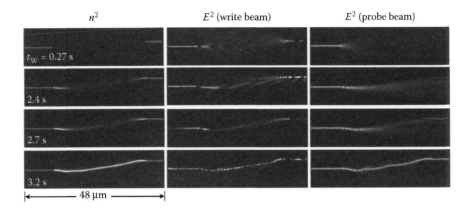

n^2 E^2 (write beam) E^2 (probe beam)

$t_W = 0.27$ s

2.4 s

2.7 s

3.2 s

|←——— 48 µm ———→|

FIGURE 4.27 Two-photon LA-SOLNET formation between 600-nm-wide waveguides with a lateral misalignment of 3000 nm.

are extremely large, the SOLNET formation accidentally fails, probably due to an unfavorable overlapping condition of write beams. To solve this problem, improved design of light source arrangements and luminescent target shapes might be required for optimization of the write-beam-propagating distribution.

Coupling Efficiency

Dependence of coupling efficiency on the writing time (Figure 4.29) and dependence of maximum coupling efficiency on lateral misalignments (Figure 4.30) in LA-SOLNETs are found to have similar tendency as in TB-SOLNETs. The lateral misalignment tolerance increases in the order of the butt joints, one-photon LA-SOLNETs, and two-photon LA-SOLNETs. The drop of the coupling efficiency observed at the 4200-nm misalignment is attributed to the accidental failure of the two-photon SOLNET formation described above.

4.2.6 PERFORMANCE OF COUPLINGS

In Figure 4.31, performance of optical couplings between 600-nm-wide waveguides is compared. Here, the coupling loss at 1000-nm lateral misalignment, which is presented in the lower figure, gives criteria for judging the applicability of the coupling method to real systems when the die bonding accuracy is assumed ∼1 µm [15].

In the butt joint, the lateral misalignment tolerance for 1-dB coupling loss and the coupling loss at 1000-nm lateral misalignment are 80 nm and 11 dB, respectively.

In the TB-SOLNET and P-SOLNET, the tolerance is widened by one order of magnitude to be 870 nm, and the coupling losses are reduced to 1.4 and 1.1 dB for the one-photon and two-photon SOLNETs, respectively. In the R-SOLNET, the tolerance is 710 nm, and the coupling loss is 2.0 and 1.2 dB for the one-photon and two-photon SOLNETs, respectively. In the LA-SOLNET, the tolerance is widened up to 2050 nm, and the coupling loss is reduced to 0.4 and 0.6 dB for the one-photon and two-photon SOLNETs, respectively.

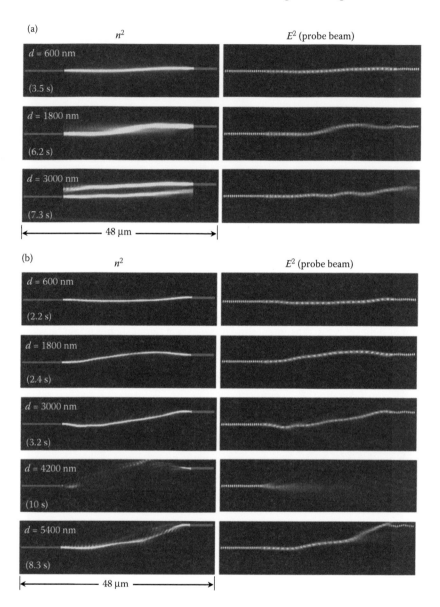

FIGURE 4.28 (a) One-photon and (b) two-photon LA-SOLNETs between 600-nm-wide waveguides at optimum writing time indicated in brackets for various lateral misalignments.

These results indicate that LA-SOLNET exhibits lateral misalignment tolerances wide enough to exceed the die bonding accuracy of ~1 μm in cases that coupling losses of 0.4–0.6 dB are permitted. This suggests that the LA-SOLNET is applicable to the optical solder for optical couplings between nanoscale waveguides. The TB-SOLNET, P-SOLNET, and R-SOLNET might also be applicable in cases that

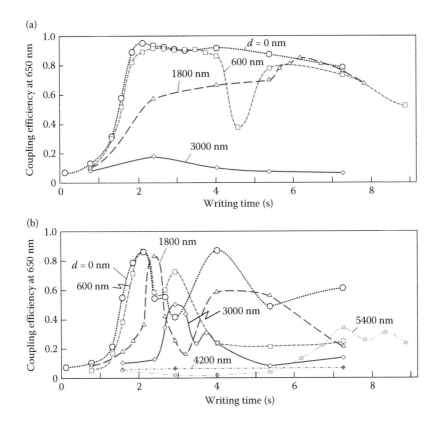

FIGURE 4.29 Dependence of coupling efficiency on writing time in LA-SOLNETs formed between 600-nm-wide waveguides. (a) One-photon LA-SOLNET and (b) Two-photon LA-SOLNET.

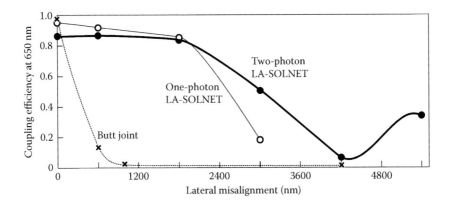

FIGURE 4.30 Dependence of maximum coupling efficiency on lateral misalignments in LA-SOLNETs formed between 600-nm-wide waveguides.

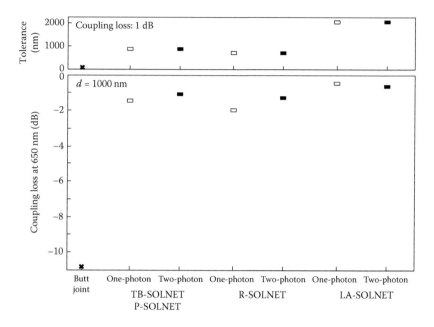

FIGURE 4.31 Performance of optical couplings between 600-nm-wide waveguides for the butt joint and SOLNETs. The upper represents lateral misalignment tolerances for 1-dB coupling loss, and the lower represents coupling losses at 1000-nm lateral misalignment.

coupling losses of 1.1–2.0 dB are permitted, allowing us to widen the choice variation for SOLNET formation processes.

4.3 SOLNETs BETWEEN MICROSCALE AND NANOSCALE WAVEGUIDES

In this section, results of simulations performed by the 2D FDTD method are presented for self-aligned optical couplings of TB-SOLNET, P-SOLNET, R-SOLNET, and LA-SOLNET between a microscale waveguide and a nanoscale waveguide [6]. The difference in the coupling performance between one-photon SOLNETs and two-photon SOLNETs is clarified.

4.3.1 SIMULATION MODELS AND PROCEDURES

Simulation Models

Models for 2D FDTD simulations of SOLNETs between a microscale waveguide and a nanoscale waveguide are shown in Figure 4.32. The widths of the input and output waveguides are respectively 3 μm and 600 nm. These optical waveguides are separated with a gap distance of 32 μm. The refractive indices of the waveguide cores, the cladding region, and the PRI material are the same as those in the models shown in Figure 4.14 described in Section 4.2.1. The lateral misalignment between the input and output waveguides is expressed by d.

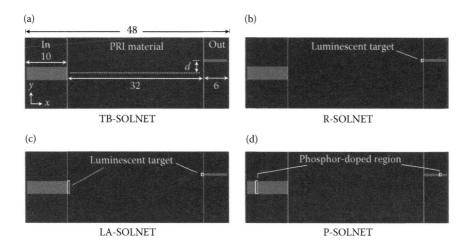

FIGURE 4.32 Models for 2D FDTD simulations of SOLNETs between 3-μm-wide microscale waveguide and 600-nm-wide nanoscale waveguide (units: μm). (a) TB-SOLNET (micro-nano), (b) R-SOLNET (micro-nano), (c) LA-SOLNET (micro-nano), and (d) P-SOLNET (micro-nano).

In the TB-SOLNET (Figure 4.32a), a λ_1-write beam is emitted from the input waveguide on the left and a λ_2-write beam from the output waveguide on the right. In the 3-μm-wide input waveguide seventy five point light sources for the λ_1-write beam are distributed across the core along y-direction with a pitch of 40 nm at $x = 1.4$ μm. In the 600-nm-wide output waveguide, thirty point light sources for the λ_2-write beam are distributed with a pitch of 20 nm at $x = 47.4$ μm. The electric field of the point light source is expressed by Equation 4.2.

In the R-SOLNET (Figure 4.32b), a 600-nm-wide line-shaped luminescent target is placed on the front edge of the output waveguide. A λ_1-write beam is introduced into the PRI material from the input waveguide, and the luminescent target emits luminescence of λ_2. The point light source arrangements in the luminescent target are the same as those for the λ_2-write beam in the TB-SOLNET. The electric-field amplitude of the emitted luminescence is obtained by the same procedure as that described in Section 4.2.1. Efficiency for the luminescence generation is assumed 0.7.

When the waveguide core has emissive characteristics intrinsically, R-SOLNETs can be formed without the luminescent target. In this case, the luminescent target in the model is regarded as the luminescent surface of the output waveguide core edge.

In the LA-SOLNET (Figure 4.32c), by excitation lights from outside, the λ_1-write beam is generated from a 3-μm-wide line-shaped luminescent target on the front edge of the input waveguide and the λ_2-write beam from a 600-nm-wide line-shaped luminescent target on the front edge of the output waveguide. The point light source arrangements in the luminescent targets on the front edges of the input and output waveguides are respectively the same as those for the λ_1-write beam and the λ_2-write beam in the TB-SOLNET.

In the P-SOLNET (Figure 4.32d), by excitation lights from outside, the λ_1-write beam and the λ_2-write beam are respectively generated from a 3-μm-wide line-shaped phosphor-doped region at $x = 1.4$ μm in the input waveguide and from a 600-nm-wide line-shaped phosphor-doped region at $x = 47.4$ μm in the output waveguide. The point light source locations and the point light source arrangements in the phosphor-doped regions are the same as those for the write beams in the TB-SOLNET. This means, as mentioned in Section 4.2.1, that the results for TB-SOLNETs are the results for both of TB-SOLNETs and P-SOLNETs.

When the waveguide core is emissive intrinsically, P-SOLNETs can be formed without the phosphor-doped regions. In this case, the phosphor-doped region in the model corresponds to the waveguide core portion, which is exposed to the excitation light.

The write-beam wavelengths are $\lambda_1 = 500$ nm and $\lambda_2 = 600$ nm. The wavelength of the probe beam, which propagates from the input waveguide to the output waveguide, is 650 nm.

Simulation Procedures

The simulation procedures for the micro–nano couplings are the same as those for the nano–nano couplings described in Section 4.2.1.

4.3.2 BUTT JOINT

Figure 4.33 shows simulation results for butt joint couplings between a 3-μm-wide waveguide and a 600-nm-wide waveguide. Considerable leakage is observed even when no lateral misalignments exist. With increasing the misalignment, the leakage increases, and for a lateral misalignment of 1800 nm, the probe-beam coupling becomes almost impossible.

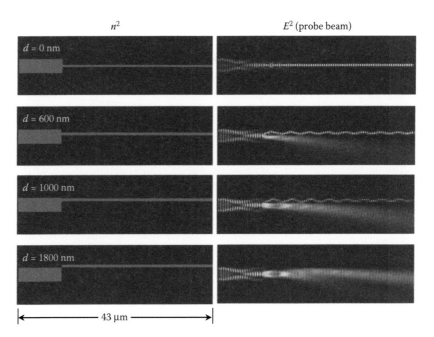

FIGURE 4.33 Probe-beam propagation in butt joints between 3-μm-wide and 600-nm-wide waveguides.

4.3.3   TB-SOLNET/P-SOLNET

SOLNET Formation

As mentioned in Section 4.3.1, the results for TB-SOLNETs described in this section mean those for both of TB-SOLNETs and P-SOLNETs.

Figure 4.34a shows one-photon TB-SOLNET formation between 3-μm-wide and 600-nm-wide waveguides with no lateral misalignments. With writing time, a coupling waveguide of TB-SOLNET is formed to guide a probe beam. However, broadening of

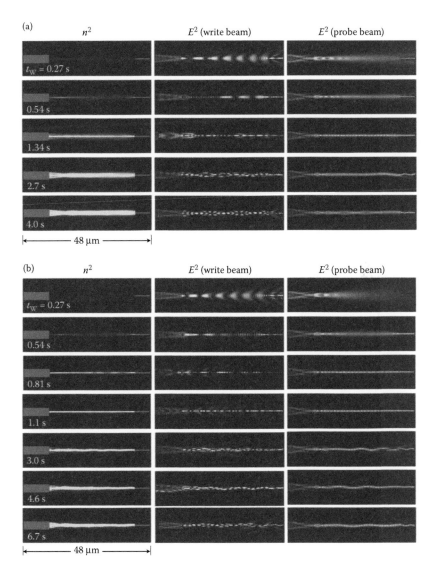

FIGURE 4.34   (a) One-photon and (b) two-photon TB-SOLNET/P-SOLNET formation between 3-μm-wide and 600-nm-wide waveguides with no lateral misalignments.

the coupling waveguide due to overexposure causes mode disturbances in the probe-beam propagation. In one-photon SOLNETs, because the rate of the refractive index increase is proportional to $\gamma_1 I_1 + \gamma_2 I_2$, the refractive index increase can develop by one write beam, namely, λ_1-write beam or λ_2-write beam. This causes unlimited broadening of the SOLNET throughout the gap region, preventing the expected tapered feature from being formed between the microscale and nanoscale waveguides.

Two-photon TB-SOLNET formation is shown in Figure 4.34b for no lateral misalignments. With writing time, a coupling waveguide of SOLNET is formed, and at 1.1 s it combines the input and output waveguides. Although the SOLNET width is close to the width of the 600-nm-wide waveguide and is much narrower than the width of the 3-μm-wide waveguide to give a straight-line-shaped coupling waveguide, the probe beam is efficiently guided from the 3-μm-wide waveguide to the 600-nm-wide waveguide. At 3.0 s, a zigzag shape appears in the SOLNET. This causes a probe beam to meander, resulting in probe-beam leakage.

After 3.0 s, broadening of the SOLNET occurs mainly on the 3-μm-wide-waveguide side. At 4.6–6.7 s a coupling waveguide with a tapered shape is formed to confine the probe beam strongly and guide it into the 600-nm-wide waveguide efficiently. In two-photon SOLNETs, because the rate of the refractive index increase is proportional to $I_1 I_2$, the refractive index increase develops by the two-write-beam overlapping. This suppresses the broadening of the SOLNET near the nanoscale waveguide, producing the taper feature.

As shown in Figure 4.35, in the one-photon TB-SOLNET the misalignment limit is ~600 nm. In the two-photon TB-SOLNET a self-aligned coupling waveguide is still formed when the misalignment is 3000 nm. The probe beam is guided into the output waveguide smoothly for lateral misalignments of 0, 600, and 1800 nm. Thus, in the micro–nano couplings similarly as in the nano–nano couplings, the two-photon TB-SOLNET is found to achieve a wider lateral misalignment tolerance comparing to the one-photon TB-SOLNET.

The probe-beam leakage occurring near the output waveguide in the two-photon TB-SOLNET with a lateral misalignment of 3000 nm is due to the small bending radius of the SOLNET.

Coupling Efficiency

Figure 4.36 shows dependence of coupling efficiency on the writing time in TB-SOLNETs formed between 3-μm-wide and 600-nm-wide waveguides. An increase in lateral misalignments causes the optimum writing time to become long and the writing time window to become narrow similarly as in the case of TB-SOLNETs between nanoscale waveguides.

For the one-photon TB-SOLNET with no lateral misalignments, the coupling efficiency rises with writing time to reach 97% (coupling loss: 0.13 dB), and then decreases. For the two-photon TB-SOLNET with no lateral misalignments, two peaks of coupling efficiency appear. Initially, the coupling efficiency rises to reach 93% (0.3 dB) at 0.81–1.6 s. Then, the efficiency decreases, and again rises to be maintained in a range of 80%–98% (1–0.1 dB) after 4.0 s. The first peak at 0.81–1.6 s is achieved by the straight-line-shaped coupling waveguide while the second peak after 4.0 s is achieved by the tapered coupling waveguide shown in Figure 4.34b.

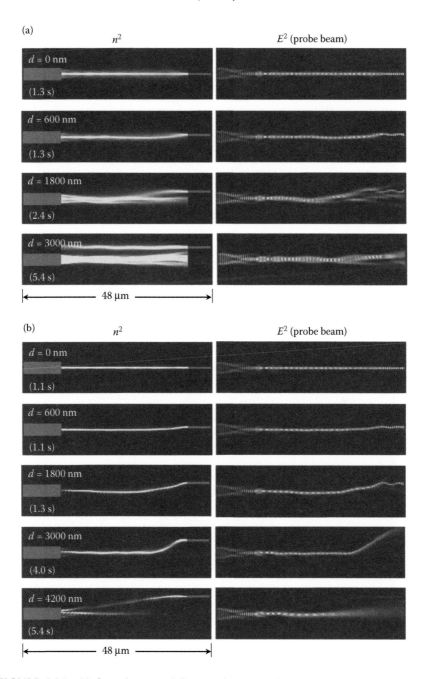

FIGURE 4.35 (a) One-photon and (b) two-photon TB-SOLNETs/P-SOLNETs between 3-μm-wide and 600-nm-wide waveguides at optimum writing time indicated in brackets for various lateral misalignments.

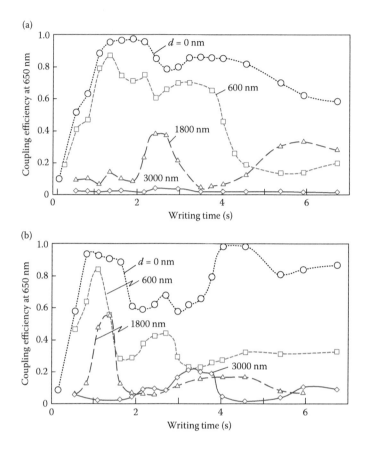

FIGURE 4.36 Dependence of coupling efficiency on writing time in TB-SOLNETs/P-SOLNETs formed between 3-μm-wide and 600-nm-wide waveguides. (a) One-photon TB-SOLNET/P-SOLNET and (b) Two-photon TB-SOLNET/P-SOLNET.

In Figure 4.37, dependence of maximum coupling efficiency on lateral misalignments is shown for TB-SOLNETs. Peak values of coupling efficiency in Figure 4.36 are plotted as a function of lateral misalignments. The coupling efficiency of the butt joint is 64% (1.9 dB) for no lateral misalignments. By introducing one-photon and two-photon TB-SOLNETs, the efficiency greatly increases to 97% (0.13 dB) and 93% (0.3 dB), respectively. In accordance with the results shown in Figure 4.35, the lateral misalignment tolerance is wider in two-photon SOLNETs than in one-photon SOLNETs.

4.3.4 R-SOLNET

SOLNET Formation

In R-SOLNETs, as Figure 4.38 shows, both in one-photon SOLNETs and in two-photon SOLNETs, coupling optical waveguides are formed for lateral misalignments of 600 and 1800 nm, and when the lateral misalignment increases to 3000 nm, no traces of SOLNET formation are observed.

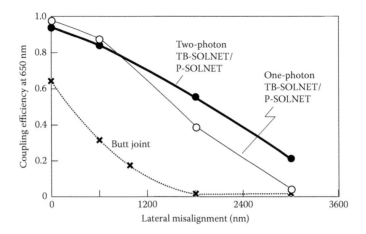

FIGURE 4.37 Dependence of maximum coupling efficiency on lateral misalignments in TB-SOLNETs/P-SOLNETs formed between 3-μm-wide and 600-nm-wide waveguides.

The limited lateral misalignment tolerances in the R-SOLNETs can be explained from a view point of the write beam and luminescence divergence in the PRI material. As can be seen in Figure 4.39, in nano–nano couplings, the spread angle of the λ_1-write beam emitted from the input waveguide is large. Consequently, even when the lateral misalignment increases to 3000 nm, the luminescent target emits strong luminescence, whose intensity is comparable to the intensity in the 1800-nm misalignment case, enabling two-photon R-SOLNET formation as shown in Figure 4.24b. On the contrary, in micro–nano couplings, because the spread angle of the λ_1-write beam is small, the luminescence becomes weak when the lateral misalignment increases to 3000 nm as shown in Figure 4.39, preventing the R-SOLNET formation.

These results suggest that the lateral misalignment tolerance of the two-photon R-SOLNET will be widened by increasing the spread angle of the λ_1-write beam emitted from the input waveguide.

Coupling Efficiency

As Figure 4.40 shows, in R-SOLNETs, there are not remarkable differences in the writing-time dependence of the coupling efficiency between one-photon and two-photon SOLNETs. Figure 4.41 shows that the lateral misalignment tolerance is almost the same for the one-photon and two-photon SOLNETs except for the irregular point at the lateral misalignment of 1800 nm. The increase in the coupling efficiency observed around writing time of 3 s in the one-photon SOLNET for a lateral misalignment of 1800 nm might be caused accidentally by the directional coupling effect.

4.3.5 LA-SOLNET

SOLNET Formation

In one-photon LA-SOLNETs, as shown in Figure 4.42, the lateral misalignment limit to form smooth coupling waveguides is ~600 nm. In two-photon LA-SOLNETs, a coupling path is still formed when the lateral misalignment increases to 3000 nm.

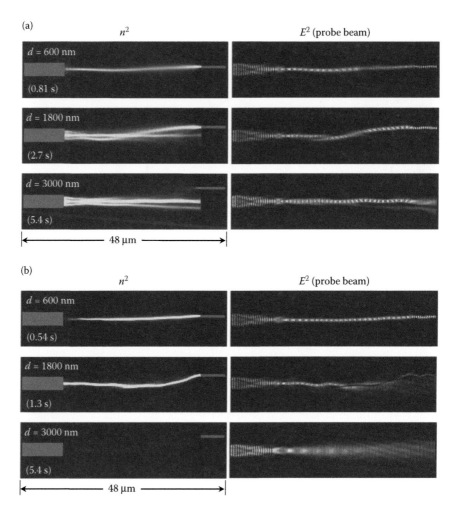

FIGURE 4.38 (a) One-photon and (b) two-photon R-SOLNETs between 3-μm-wide and 600-nm-wide waveguides at optimum writing time indicated in brackets for various lateral misalignments.

One concern in the two-photon LA-SOLNETs is the branching structures that appear in SOLNETs for lateral misalignments of 1800 and 3000 nm. The branching might be attributed to the wavefront shape of the luminescence. As can be seen in Figure 4.43, in nano–nano couplings, luminescence with smooth wavefronts is emitted from the nanoscale luminescent targets, propagating in a manner of a light emitted from a single-mode optical fiber. Consequently, smooth two-photon LA-SOLNETs without branching structures are available as shown in Figure 4.28b. On the contrary, in micro–nano couplings, luminescence with a complicated wavefront is emitted from the microscale luminescent target, propagating in a plurality of directions in a manner of a light emitted from a multimode optical fiber. This might cause the branching structures of the two-photon LA-SOLNETs. It is suggested from these

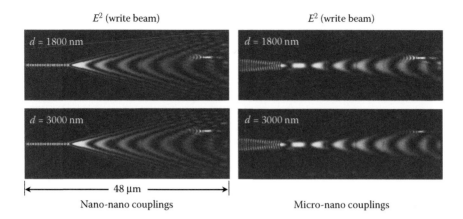

E^2 (write beam) E^2 (write beam)

$d = 1800$ nm $d = 1800$ nm

$d = 3000$ nm $d = 3000$ nm

|← 48 μm →|

Nano-nano couplings   Micro-nano couplings

FIGURE 4.39   Propagation of λ_1-write beam and luminescence emitted from the luminescent target for two-photon R-SOLNET formation at writing time of 0.27 s.

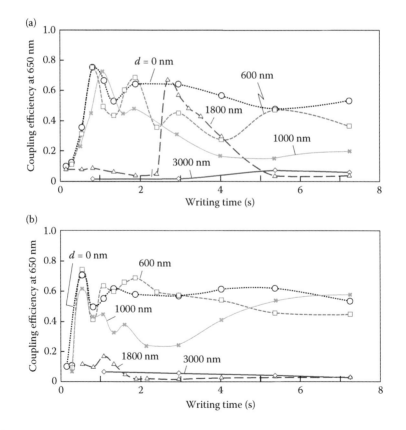

FIGURE 4.40   Dependence of coupling efficiency on writing time in R-SOLNETs formed between 3-μm-wide and 600-nm-wide waveguides. (a) One-photon R-SOLNET and (b) Two-photon R-SOLNET.

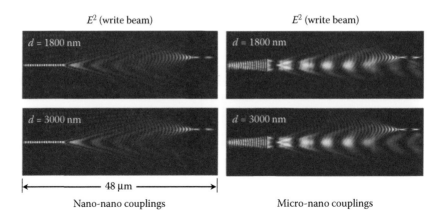

Nano-nano couplings   Micro-nano couplings

FIGURE 4.43 Propagation of luminescence emitted from the luminescent targets for two-photon LA-SOLNET formation at writing time of 0.27 s.

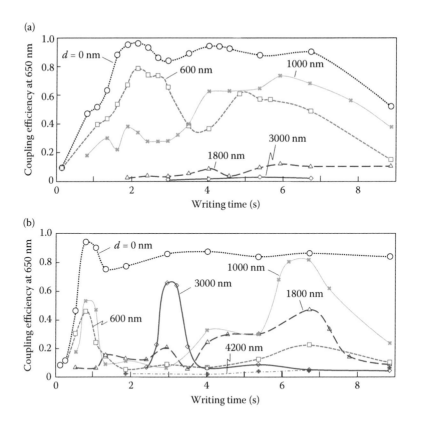

FIGURE 4.44 Dependence of coupling efficiency on writing time in LA-SOLNETs formed between 3-μm-wide and 600-nm-wide waveguides. (a) One-photon LA-SOLNET and (b) Two-photon LA-SOLNET.

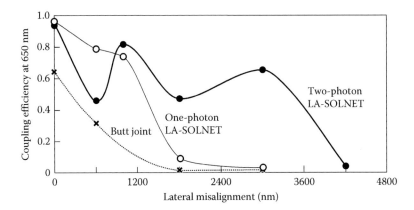

FIGURE 4.45 Dependence of maximum coupling efficiency on lateral misalignments in LA-SOLNETs formed between 3-μm-wide and 600-nm-wide waveguides.

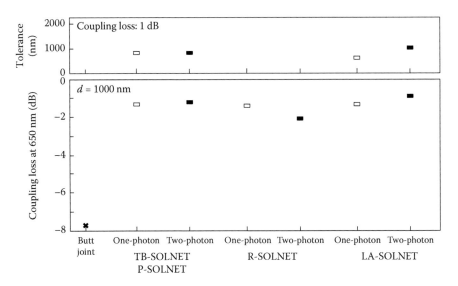

FIGURE 4.46 Performance of optical couplings between 3-μm-wide and 600-nm-wide waveguides for the butt joint and SOLNETs. The upper represents lateral misalignment tolerances for 1-dB coupling loss, and the lower represents coupling losses at 1000-nm lateral misalignment.

loss of 1.2 dB is permitted. R-SOLNETs might also be applicable when a coupling loss of 1.4 dB is permitted.

4.4 INFLUENCE OF WRITE-BEAM WAVELENGTHS ON SOLNET FORMATION

In this section, results for TB-SOLNETs (at the same time for P-SOLNETs as mentioned in Section 4.2.1) formed by a write beam combination of [$\lambda_1 = 400$ nm,

$\lambda_2 = 780$ nm] are presented [16], and they are compared with results for TB-SOLNETs formed by a write beam combination of [$\lambda_1 = 500$ nm, $\lambda_2 = 600$ nm], which are described in Section 4.2.3. The simulation model is the same as that depicted in Figure 4.14a, and the 2D FDTD simulation procedures are the same as those described in Section 4.2.1.

Figure 4.47 shows TB-SOLNETs formed by $\lambda_1 = 400$ nm and $\lambda_2 = 780$ nm for various lateral misalignments. In the one-photon SOLNET, self-aligned

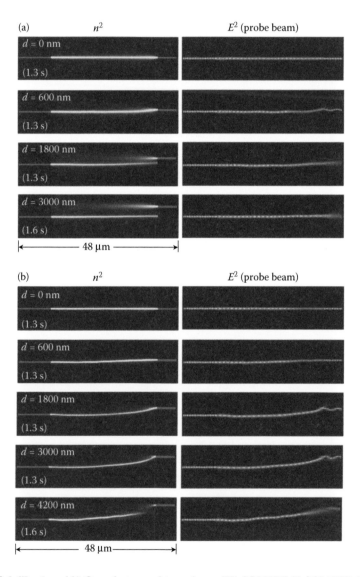

FIGURE 4.47 (a and b) One-photon and two-photon TB-SOLNETs/P-SOLNETs between 600-nm-wide waveguides at optimum writing time indicated in brackets for various lateral misalignments [$\lambda_1 = 400$ nm and $\lambda_2 = 780$ nm].

coupling waveguides are formed between the input and output waveguides for lateral misalignments of 0 and 600 nm. Probe beams are concentrated into the output waveguides. The coupling efficiency is 85% for 600-nm lateral misalignment. When the lateral misalignment is increased to 1800 nm or more, self-aligned coupling waveguides are unable to form. Two optical waveguides extending from the input and output waveguides are completely separated to give low coupling efficiency of 19% for 1800-nm misalignment and 3% for 3000-nm misalignment.

In the two-photon SOLNET, even when the lateral misalignment is increased to 1800 nm, a self-aligned coupling waveguide is formed with coupling efficiency of 70%. When the lateral misalignment is further increased to 3000 nm, a self-aligned coupling waveguide is still formed with coupling efficiency of 60%. Thus, it is found that the two-photon TB-SOLNET drastically extends the lateral misalignment tolerance to 3000 nm, which is five times larger than the misalignment limit of ∼600 nm in the one-photon TB-SOLNET.

Figure 4.48 shows n^2 profiles across TB-SOLNETs (along y-direction) at 36 μm for 600-nm lateral misalignment. The position $y = 16$ μm corresponds to the core axis location of the input waveguide. Graded-index-like profiles are observed at the initial stage of SOLNET formation, and later change to step-index-like profiles similarly as

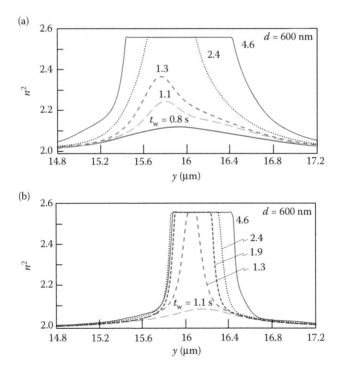

FIGURE 4.48 (a and b) Cross-sectional n^2 profiles of one-photon and two-photon TB-SOLNETs/P-SOLNETs at $x = 36$ μm [$\lambda_1 = 400$ nm and $\lambda_2 = 780$ nm].

in the TB-SOLNETs formed by write beams of $\lambda_1 = 500$ nm and $\lambda_2 = 600$ nm (see Figure 4.19). The waveguide broadening is suppressed in the two-photon SOLNET compared with that in the one-photon SOLNET.

The n^2 profiles across the SOLNETs were found to vary along the x-axis with fluctuations of the SOLNET width. At long writing time, the fluctuations become large due to overexposure of write beams to increase the propagation loss and reduce the coupling efficiency.

As can be seen in Figure 4.47b, the line shapes of the two-photon TB-SOLNETs formed by $\lambda_1 = 400$ nm and $\lambda_2 = 780$ nm are not symmetric. The two-photon SOLNETs formed by $\lambda_1 = 500$ nm and $\lambda_2 = 600$ nm, on the other hand, have symmetric S-shapes as described in Section 4.2.3 (see Figure 4.20b). The difference in the SOLNET shape is attributed to the difference in the spread angles between the λ_1-write beam and the λ_2-write beam.

The spread angle θ of write beams with a wavelength λ in a PRI material with a refractive index n is approximately estimated by the following expression.

$$\theta = \frac{\lambda}{2\pi n w} \tag{4.7}$$

Here, w is the waveguide core width. The spread angles in a medium with a refractive index of 1.5 are calculated to be 0.15 and 0.29 rad for [$\lambda_1 = 400$ nm, $\lambda_2 = 780$ nm] and 0.18 and 0.22 rad for [$\lambda_1 = 500$ nm, $\lambda_2 = 600$ nm]. In the latter case, the spread angle difference between the λ_1-write beam and the λ_2-write beam is much smaller than in the former case, resulting in the symmetric S-shaped two-photon SOLNETs.

As described in Section 4.2.3, the symmetric S-shaped structure of the two-photon TB-SOLNETs formed by $\lambda_1 = 500$ nm and $\lambda_2 = 600$ nm gives rise to a wide lateral misalignment tolerance of 4200 nm, which is 1.4-times larger than the tolerance of 3000 nm in the two-photon TB-SOLNETs formed by $\lambda_1 = 400$ nm and $\lambda_2 = 780$ nm. A comparison of the results in Figures 4.20b and 4.47b reveals that the probe-beam meandering is suppressed in the two-photon SOLNETs formed by $\lambda_1 = 500$ nm and $\lambda_2 = 600$ nm, meaning a reduction in the mode disturbances of the probe beam in the output waveguide.

These results suggest that by adjusting the write beam wavelengths as well as waveguide core widths and core edge structures so that the spread angles of the λ_1-write beam and the λ_2-write beam become as close as possible, SOLNETs with higher coupling efficiency are expected to be obtained.

4.5 INFLUENCE OF WRITE-BEAM INTENSITY AND GAP DISTANCES ON SOLNET FORMATION

As mentioned in Section 4.2.3, one concern in SOLNETs, particularly in two-photon SOLNETs, is the rapid decrease in the coupling efficiency with writing time after the maximum has been reached to narrow the writing-time windows. Another concern is the lateral misalignment dependence of the optimum writing time required to reach the maximum coupling efficiency.

For practical use of the SOLNET optical solder, it is desirable that the writing time window is wide and the optimum writing time is constant regardless of the lateral misalignment to form SOLNETs at fixed writing time for the unmonitored optical solder; otherwise, we would have to monitor the coupling efficiency and adjust the writing time one by one to obtain the maximum efficiency.

One of the approaches to widen the writing time window might be a slowdown in the SOLNET formation rate by reducing the write-beam intensity. To suppress the lateral misalignment dependence of the optimum writing time, the adjustment of the gap distance between input and output waveguides might be a possible approach.

In this section, influence of the write-beam intensity and the gap distance on SOLNET formation are described based on the results of 2D FDTD simulations.

4.5.1 WRITE-BEAM INTENSITY

In two-photon SOLNETs, when the lateral misalignment is large, the coupling efficiency between input and output waveguides rapidly decreases with writing time as can be seen in Figure 4.21b. This might be caused by the zigzag SOLNET structures that appear after long exposure to the write beams (see n^2 distribution at 5.4 s in Figure 4.18b). The zigzag structures are believed to be a consequence of the steep formation characteristics of the two-photon SOLNET that can be seen in Figure 4.19. So, it is expected that by reducing the write-beam intensity the rapid coupling efficiency decrease is suppressed to widen the writing time windows, which is favorable for the unmonitored optical solder.

Figure 4.49 shows influence of the write-beam intensity on the coupling efficiency versus writing time characteristics in two-photon TB-SOLNETs formed between 600-nm-wide waveguides with a lateral misalignment of 3000 nm by $\lambda_1 = 500$ nm and $\lambda_2 = 600$ nm. The simulation model is the same as that depicted in Figure 4.14a.

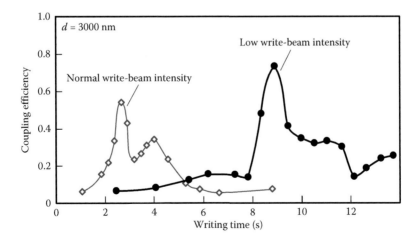

FIGURE 4.49 Influence of write-beam intensity on the coupling efficiency vs. writing-time characteristics in two-photon TB-SOLNETs/P-SOLNETs formed between 600-nm-wide waveguides with a lateral misalignment of 3000 nm.

In the figure, "normal write-beam intensity" means the intensity used in the simulation for Figure 4.21b described in Section 4.2.3, and "low write-beam intensity" indicates that the intensity is reduced to one-half the normal write-beam intensity. It is found that the maximum coupling efficiency increases and the writing time window tends to be widened by forming a SOLNET at a slower rate. Further slowdown in the SOLNET formation rate might enhance the tendency of the writing time window widening.

4.5.2 Gap Distances

To realize unmonitored optical solder of SOLNETs, suppression of the lateral misalignment dependence of the optimum writing time is desirable. Controlling the gap distance between input and output waveguides is one of the approaches to solve the issue [11].

Figure 4.50 shows the model used for simulations to investigate the influence of gap distances on the optimum writing time. Calculations were performed for one-photon R-SOLNETs with luminescent targets. The simulation model is almost similar to that shown in Figure 4.32b. The widths of input and output waveguides are respectively 3 µm and 600 nm. The gap distance L is varied from 8 to 64 µm. The refractive indices of the input waveguide, the output waveguide, and the cladding region are 1.52, 2.0, and 1.5, respectively. Wavelengths of the write beam from the input waveguide, the luminescence, and the probe beam from the input waveguide are respectively 470, 530, and 650 nm. The efficiency for luminescence generation is assumed 0.2.

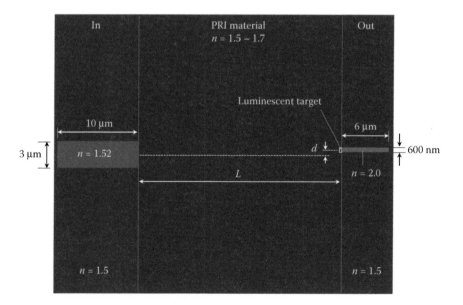

FIGURE 4.50 Model for 2D FDTD simulations to investigate the influence of gap distances on one-photon R-SOLNET formation. (From T. Yoshimura, *J. Lightwave Technol.* **33**, 849–856, 2015.) [11]

Figure 4.51 shows simulation results for butt joint couplings. Considerable leakages caused by the large core width mismatch between the two waveguides are observed even when no misalignments exist. By increasing the lateral misalignment, the leakage increases, and at the same time the probe-beam meandering in the output waveguide becomes remarkable. The coupling efficiency is 36% and 29% for the lateral misalignments of 0 and 600 nm, respectively.

Figure 4.52 shows one-photon R-SOLNET formation for a gap distance of 32 μm. For the case that no lateral misalignments exist, as shown in Figure 4.52a, when a write beam is introduced from the input waveguide into the PRI material, R-SOLNET is gradually formed. Accordingly, the probe beam, which initially diffuses, is gradually concentrated into the output waveguide with writing time. At 5.4 s, however, probe-beam leakage is observed at the joint between the 3-μm-wide input waveguide and the R-SOLNET, which is caused by overly-strong probe-beam confinement arising from overexposure.

When a lateral misalignment of 600 nm exists between the waveguides, as shown in Figure 4.52b, an R-SOLNET is formed toward the output waveguide because of the "pulling water" effect of the luminescent target, and the probe beam is guided accordingly into the output waveguide. When the lateral misalignment is increased up to 900 nm, as shown in Figure 4.52c, the targeting effect is weakened, and this induces considerable probe-beam leakage at the joint between the R-SOLNET and the 600-nm-wide output waveguide.

In Figure 4.53, the gap distance dependence of the one-photon R-SOLNET formation and the probe-beam propagation is shown for a lateral misalignment of 600 nm at writing time of 5.4 s. When the gap distance is 8 μm, considerable probe-beam leakage is observed at the front edge of the output waveguide. For gap distances of more than 16 μm, smooth guiding of the probe beam from the input waveguide into the output waveguide is observed, indicating that self-aligned optical couplings have been established.

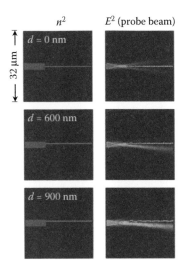

FIGURE 4.51 Probe-beam propagation in butt joints between 3-μm-wide and 600-nm-wide waveguides.

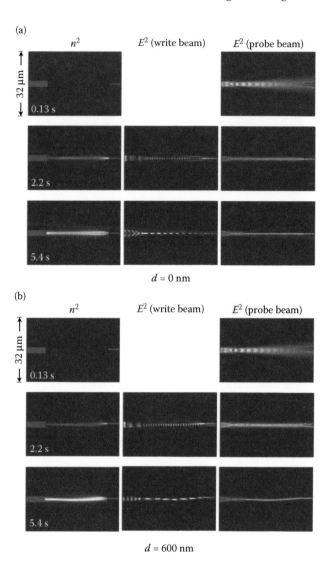

FIGURE 4.52 One-photon R-SOLNET formation between 3-μm-wide and 600-nm-wide waveguides for a gap distance of 32 μm.   *(Continued)*

Figures 4.54a–c show the coupling efficiency of the probe beam from the input waveguide to the output waveguide as a function of the writing time for gap distances of 16, 32, and 64 μm. For lateral misalignments of 0 and 600 nm, regardless of the gap distance, the coupling efficiency increases with writing time to reach a maximum, and then decreases to a steady-state level. This coupling efficiency decrease is attributed to overexposure, which induces the overly-strong probe-beam confinement, resulting in leakage, as can typically be seen in the probe-beam propagation at 5.4 s in Figure 4.52a. The optimum writing time required to reach the maximum coupling efficiency tends to increase with increasing the lateral misalignment.

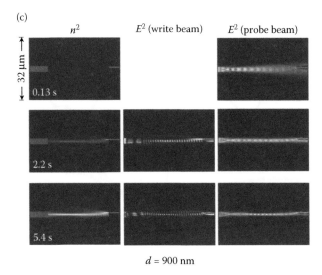

d = 900 nm

FIGURE 4.52 (Continued) One-photon R-SOLNET formation between 3-μm-wide and 600-nm-wide waveguides for a gap distance of 32 μm.

For a lateral misalignment of 900 nm, the coupling efficiency is low when compared with that for lateral misalignments of 0 and 600 nm, indicating that the R-SOLNET effect is weakened.

In Figure 4.55, peak values of coupling efficiency in Figure 4.54 are plotted as a function of the gap distance for a lateral misalignment of 600 nm. It is found

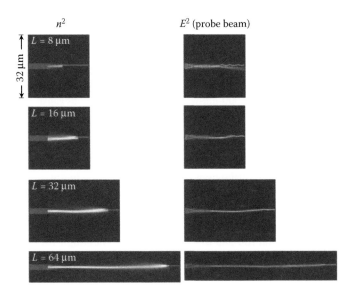

FIGURE 4.53 One-photon R-SOLNETs between 3-μm-wide and 600-nm-wide waveguides for various gap distances at writing time of 5.4 s. The lateral misalignment is 600 nm.

(a)

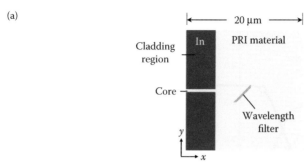

Model for L-shaped SWW formation

(b) E^2 (write beam)

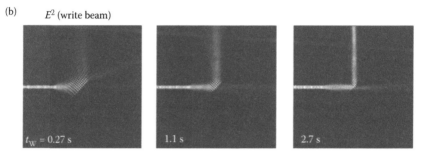

L-shaped SWW formation via wavelength filter

FIGURE 4.56   (a) Model for 2D FDTD simulations of L-Shaped SWWs. (b) L-Shaped SWW formation from a 500-nm-wide waveguide via a wavelength filter. (From H. Kaburagi and T. Yoshimura, *Opt. Commun.* **281**, 4019–4022, 2008.) [17]

diffraction. Then, the write beam gradually exhibits self-focusing. Finally, an L-shaped SWW is formed along the center axis of the write-beam propagation.

4.6.2   R-SOLNET with Wavelength Filters

Formation of R-SOLNETs with wavelength filters, which are used in the simulation of vertical R-SOLNETs in Section 4.6.3, is simulated by the 2D FDTD method [17]. Figure 4.57 shows simulation models for couplings between a 2-μm-wide input waveguide and a 500-nm-wide output waveguide with a lateral misalignment of 500 nm. The refractive indices of the core and cladding region are 1.5 and 1.0, respectively. The refractive index of the PRI material increases from 1.5 to 1.8 upon write-beam exposure. Figure 4.57a is a model for a SWW, and Figure 4.57b is a model for an R-SOLNET, where a dielectric multilayer wavelength filter is placed on the front edge of the output waveguide. The wavelength filter, which reflects 650-nm write beams and transmits 1.3-μm probe beams, consists of 80-nm-thick layer with refractive index of 1.5 and 80-nm-thick layer with refractive index of 2.5. Total layer count is four.

The simulation procedure is the same as that for the SOLNETs between nanoscale waveguides described in Section 4.2.1. A write beam, which is generated

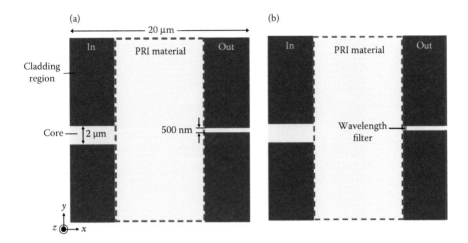

FIGURE 4.57 Models for 2D FDTD simulations of (a) SWWs and (b) R-SOLNETs with wavelength filters between 2-μm-wide and 500-nm-wide waveguides with a lateral misalignment of 500 nm. (From H. Kaburagi and T. Yoshimura, *Opt. Commun.* **281**, 4019–4022, 2008.) [17]

from point light sources located at the left edge of the input waveguide with a pitch of 20 nm, is introduced from the input waveguide into the PRI material. The point light source locations and arrangements for the probe beam are the same as those for the write beam.

For the SWW, as shown in Figure 4.58a, when a write beam is introduced from the input waveguide into the PRI material, the refractive index increases with writing time along the center axis of the write-beam propagation path to form a straight waveguide of a SWW. Due to the lateral misalignment of 500 nm, misalignment is caused between the SWW and the output waveguide. For the R-SOLNET, on the other hand, as shown in Figure 4.58b, a waveguide is formed toward the location of the wavelength filter. Namely, a part of the write beam introduced from the input waveguide is reflected by the wavelength filter, so that the reflected write beam overlaps the incident write beam. This induces the "pulling water" effect to construct a self-aligned coupling waveguide of R-SOLNET between the input and output waveguides.

Figure 4.59a shows propagation of a probe beam introduced from the input waveguide for the SWW. Before SWW formation, due to diffraction, the probe beam expands with propagation, resulting in a low coupling efficiency of 28%. After SWW formation, the probe beam is confined into the SWW to be guided to the output waveguide with a 2.5-times increase in the coupling efficiency. However, due to the misalignment between the SWW and the output waveguide, a considerable leakage is induced at the joint. Furthermore, higher-order-mode probe-beam propagation is induced in the output waveguide, resulting in probe-beam meandering.

In the case of the R-SOLNET, as Figure 4.59b shows, before SOLNET formation, coupling efficiency is 20%, which is slightly lower compared to the efficiency in the SWW. This might be attributed to loss at the wavelength filter on the front edge of

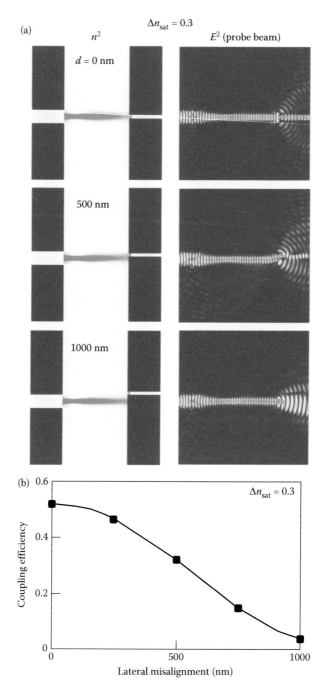

FIGURE 4.60 (a) R-SOLNET formation with wavelength filters between 2-μm-wide and 500-nm-wide waveguides and (b) dependence of coupling efficiency on lateral misalignments in a PRI material with $\Delta n_{sat} = 0.3$ at writing time of 4.8 s.

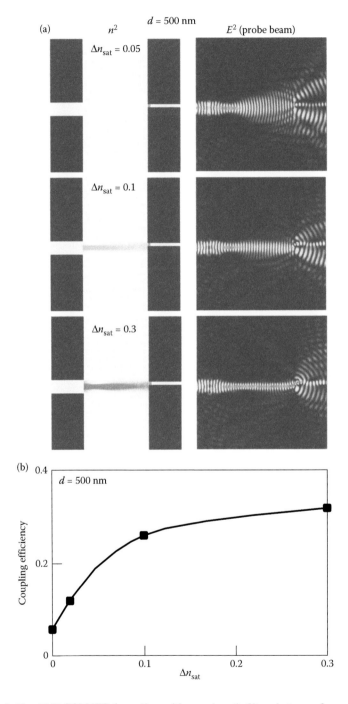

FIGURE 4.61 (a) R-SOLNET formation with wavelength filters between 2-μm-wide and 500-nm-wide waveguides and (b) dependence of coupling efficiency on Δn_{sat} for a lateral misalignment of 500 nm at writing time of 4.8 s.

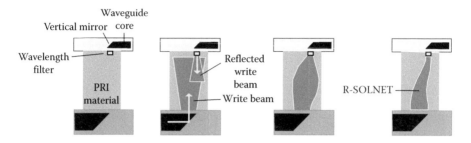

FIGURE 4.62 Process of vertical R-SOLNET formation for self-aligned optical Z-connections.

A process of vertical R-SOLNET formation to construct a self-aligned optical Z-connection is depicted in Figure 4.62. Two optical waveguides are stacked with a gap filled with a PRI material. A wavelength filter is deposited at a vertical mirror location of one of the optical waveguides. The wavelength filter reflects a write beam for the R-SOLNET formation and transmits a probe beam (or a signal beam), therefore acting as a mirror for the write beam and as a transparent object for the probe beam (or the signal beam). When a write beam from the other optical waveguide enters the free space filled with the PRI material, a part of the write beam is reflected by the wavelength filter. The incident write beam and the reflected write beam overlap. In the overlap region, the refractive index of the PRI material increases rapidly, guiding the write beam to the wavelength filter location gradually through the "pulling water" effect. Finally, the incident write beam and the reflected write beam merge to form one lightwave path by self-focusing, completing a self-aligned optical Z-connection between the two waveguides automatically.

Figure 4.63 shows a model for 2D FDTD simulations of an optical Z-connection consisting of a vertical waveguide of R-SOLNET with wavelength filters. A 1.2-μm-thick core with a 45° total internal reflection (TIR) mirror is positioned on a 300 nm-thick under-cladding layer to form an optical waveguide film. Two optical waveguide films are stacked with a 10-μm gap filled with a PRI material. A wavelength filter is placed at the TIR mirror location of the upper optical waveguide film. The refractive indices of the core (n_{Core}), under-cladding layer (n_{Clad}), and surrounding air region (n_{Air}) are 1.8, 1.5, and 1.0, respectively. The refractive index of the PRI material (n_{PRI}) varies from 1.5 to 1.7 upon write-beam exposure. The wavelengths of the write beam and the probe beam are respectively 650 and 850 nm. Point light sources are distributed with a pitch of 20 nm in the input waveguide at $x = 1.0$ μm. The polarization direction is along the z-axis.

Effects of lateral misalignments on the vertical R-SOLNET formation for optical Z-connections are shown in Figure 4.64. η in the figure is the coupling efficiency of probe beams between the input and output optical waveguides. Negative and positive lateral misalignments respectively represent left-hand-side and right-hand-side dislocation of the upper optical waveguide film from the just-aligned position. As can be seen from Figure 4.64a, when the lateral misalignment falls between 0 and 750 nm, the R-SOLNETs are led to the wavelength filter location by the "pulling water" effect. For misalignments of −250, −120, and 1000 nm, the R-SOLNETs

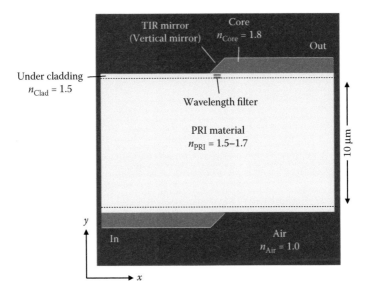

FIGURE 4.63 Model for 2D FDTD simulations of optical Z-connections consisting of vertical waveguides of R-SOLNET with wavelength filters. (From T. Yoshimura et al. *IEEE J. Select. Topics in Quantum Electron.* **17**, 566–570, 2011.) [18]

are only partially led to the wavelength filter. According to the refractive index distributions shown in Figure 4.64a, relatively low-levels of probe-beam leakages are observed at the upper TIR mirror for lateral misalignments of 0–750 nm as can be seen in Figure 4.64b.

It is noted that the R-SOLNETs depicted in Figure 4.64a exhibit left-hand-side bending toward the TIR mirror in the upper optical waveguide film when the lateral misalignment is 0 nm. This is attributed to the tunnel effect of lightwaves, namely, the effect of an evanescent wave, at the TIR mirror surface in the lower optical waveguide film.

As shown in Figure 4.65a, when a TIR mirror is used for the vertical mirror, that is, in the case of "without perfect conductor," the vertical waveguide location shifts toward the right, away from the vertical mirror location by ~200 nm due to the tunnel effect. The depth of the tunneling, d_{tunnel}, is approximately estimated by the following expression for plane waves,

$$d_{tunnel} = \frac{n_{Air}}{k_0 \sqrt{n_{Core}^2 \sin^2 \theta - n_{Air}^2}} \tag{4.8}$$

Here, k_0 is the wavenumber in a vacuum and θ is the incident angle of the light beams to the TIR mirror surface. For a wavelength of 650 nm, an incident angle of 45°, $n_{Core} = 1.8$, and $n_{Air} = 1.0$, $d_{tunnel} = 130$ nm is obtained. This value is close to the size of the ~200-nm shift that appears in the simulation. When the mirror surface is covered with a perfect conductor as shown in Figure 4.65b, the tunneling is suppressed to form a vertical waveguide just above the vertical mirror, increasing the

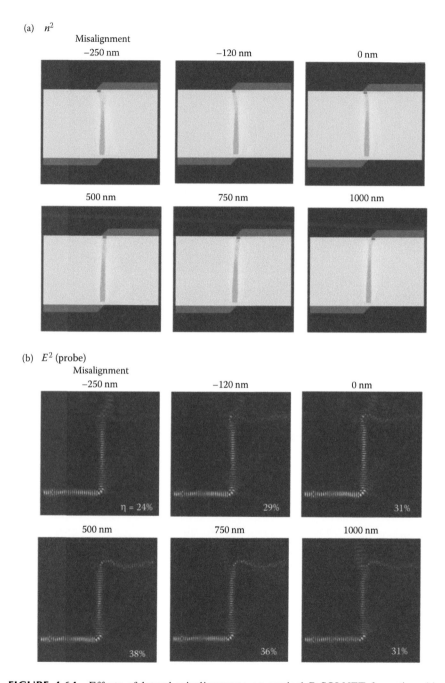

FIGURE 4.64 Effects of lateral misalignments on vertical R-SOLNET formation. (a) Refractive index distributions and (b) intensity of propagated probe beams. Writing time is 2.7 s. Negative and positive lateral misalignments respectively represent left-hand-side and right-hand-side dislocation of upper optical waveguides. (From T. Yoshimura et al. *IEEE J. Select. Topics in Quantum Electron.* **17**, 566–570, 2011.) [18]

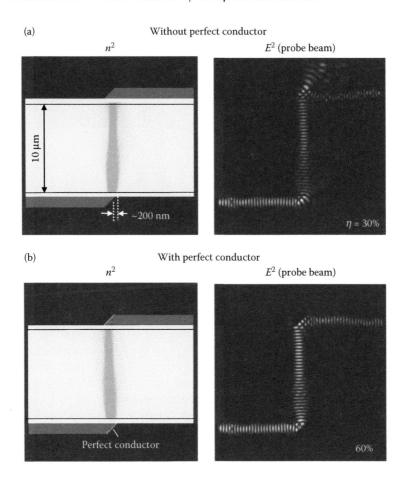

FIGURE 4.65 Effects of the perfect conductor on the position of the probe-beam reflection at vertical mirrors. (From T. Yoshimura et al. *IEEE J. Select. Topics in Quantum Electron.* **17**, 566–570, 2011.) [18]

coupling efficiency between the lower and upper optical waveguides. The coupling efficiency for the model without a perfect conductor is 30% while that for the model with a perfect conductor is 60%.

It is also noted that, in Figure 4.64b, beam leakage at the upper TIR mirror is larger in the left-hand-side-bending R-SOLNETs than in the right-hand-side-bending R-SOLNETs. This result can be explained based on the model shown in Figure 4.66. When a guided beam in an R-SOLNET propagates toward the left, the incident angle to the mirror surface becomes small, reducing the beam reflection and increasing beam leakage. When a guided beam in an R-SOLNET propagates toward the right, on the other hand, the incident angle to the mirror surface becomes large, enhancing the beam reflection and decreasing beam leakage. When a perfect conductor is deposited on the mirror surface of the upper optical waveguide film, as shown in Figure 4.67, the leakage at the mirror surface in the left-hand-side-bending R-SOLNET is drastically suppressed.

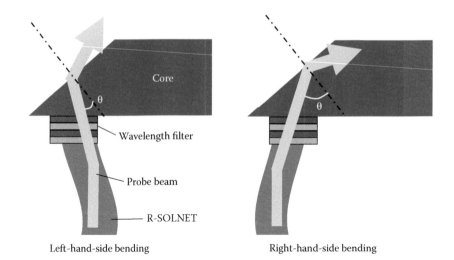

Left-hand-side bending   Right-hand-side bending

FIGURE 4.66   Models for leakage at vertical mirrors in optical Z-connections of R-SOLNET. (From T. Yoshimura et al. *IEEE J. Select. Topics in Quantum Electron.* **17**, 566–570, 2011.) [18]

Figure 4.68 summarizes the effects of lateral misalignments on coupling efficiency in optical Z-connections of R-SOLNET with wavelength filters. By putting a wavelength filter (WF) at the TIR vertical mirror location in the upper optical waveguide film ([b]), which is to say by creating an R-SOLNET, the lateral misalignment tolerance increases comparing to the case without the wavelength filter ([a]). When a perfect conductor (PC) is deposited on the vertical mirror surface in the upper optical waveguide film ([c]), the coupling efficiency increases comparing to the case of [a]. When both of a wavelength filter and a perfect conductor are implemented simultaneously ([d]), a drastic increase in coupling efficiency is observed, especially in the negative lateral misalignment region. This gives rise to a considerable increase in the system's lateral misalignment tolerance: in fact, a coupling efficiency of >40% can be kept under a condition of lateral misalignments from −200 to 800 nm in waveguide assembly. If the slight loss occurring at the wavelength filter is reduced, more improvement of the coupling efficiency would be expected.

In the models considered in this section, optical waveguides to be coupled have the same core size. Because SOLNETs are applied to optical couplings between optical waveguides with different core sizes as mentioned in Section 4.3, R-SOLNETs are expected to be effective for optical Z-connections with different core sizes, such as the structure shown in Figure 4.62.

4.7   PARALLEL SOLNETs

In advanced OE systems, parallel optical wiring is implemented. In this section, formation of parallel waveguides of R-SOLNET with luminescent targets is demonstrated by the 2D FDTD method [10]. Since calculations were performed only for one-photon SOLNETs, expressions of "one-photon" and "two-photon" are not marked.

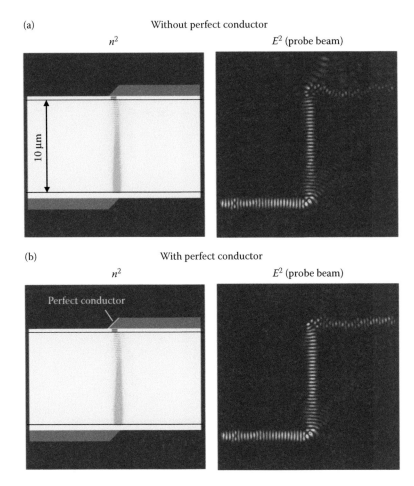

FIGURE 4.67 Effects of the perfect conductor on leakage at vertical mirrors in optical Z-connections of R-SOLNET. Writing time is 2.7 s. (From T. Yoshimura et al. *IEEE J. Select. Topics in Quantum Electron.* **17**, 566–570, 2011.) [18]

4.7.1 Comparison of Parallel R-SOLNETs and SWWs

Figure 4.69 shows a simulation model for parallel R-SOLNETs with luminescent targets. Two 600-nm-wide input waveguides are placed with a waveguide distance of P. The refractive index of the cladding region is 1.0 and that of the waveguide core is 1.8. Two 600-nm-wide luminescent targets are located with a lateral misalignment of d from the axis of the input waveguides. Between the front edges of the input waveguides and the luminescent targets, there is a 13-μm-long gap filled with a PRI material. The refractive index of the PRI material increases from 1.5 to 1.7 upon write-beam exposure.

The simulation procedures based on the 2D FDTD method are similar to those explained in Section 4.2.1. Wavelengths of the write beam, luminescence, and probe beam are respectively 400, 500, and 650 nm. In order to investigate the effect of

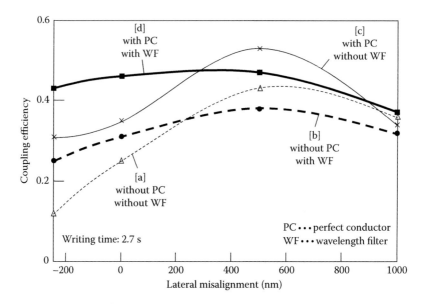

FIGURE 4.68 Dependence of coupling efficiency on lateral misalignments in optical Z-connections of R-SOLNET with wavelength filters. (From T. Yoshimura et al. *IEEE J. Select. Topics in Quantum Electron.* **17**, 566–570, 2011.) [18]

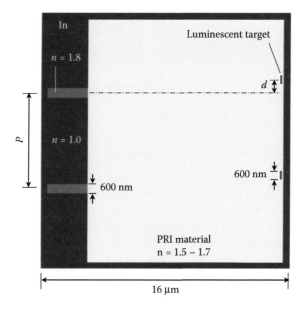

FIGURE 4.69 Model for 2D FDTD simulations of parallel R-SOLNETs with luminescent targets. (From T. Yoshimura and M. Seki, *J. Opt. Soc. Am. B* **30**, 1643–1650, 2013.) [10]

the divergence of write beams, results of simulations with a write-beam wavelength of 650 nm, a luminescence wavelength of 700 nm, and a probe-beam wavelength of 800 nm are also presented. Efficiency for the luminescence generation from the luminescent targets is assumed 0.2. The polarization direction is perpendicular to the calculated plane.

For simulations of parallel SWWs, the luminescent targets are removed in the model shown in Figure 4.69.

Figure 4.70 shows simulation results of parallel SWW formation for a write-beam wavelength of 400 nm. For a waveguide distance of 4 μm, parallel SWWs are gradually grown straight from the input waveguides. Accordingly, a 650-nm probe beam introduced from the upper input waveguide is gradually confined in the SWW with writing time. When the waveguide distance is reduced to 2 μm, the two SWWs attract each other to be bent. Accordingly, a probe beam propagates with a bow-shaped trace to deviate from the right destination. Here, the right destination means the 600-nm-wide region with its center on the axis of the corresponding input waveguide. When the waveguide distance is reduced to 1 μm, the two SWWs merge.

Figure 4.71 shows coupling efficiency of a probe beam from the upper input waveguide to the right destination. Reflecting the shapes of SWWs shown in Figure 4.70, the coupling efficiency increases with writing time up to 80% for a waveguide distance of 4 μm while the efficiency for 2 and 1 μm is limited around 20%.

As can be seen in Figure 4.70b, the parallel SWWs deviate from the right destinations for a waveguide distance of 2 μm. This problem can be solved by using R-SOLNETs with luminescent targets as shown in Figure 4.72. Luminescent targets are placed at the right destinations, where the SOLNET should be drawn, namely, each target is placed on the axis of the corresponding input waveguide. The luminescence induced by the write beams pulls the optical waveguides of R-SOLNET back to the right destinations. A probe beam is guided toward the right destination along the R-SOLNET.

In Figure 4.73, coupling efficiency of a probe beam to the right destination through the optical waveguide of R-SOLNET is compared with that through the SWW for a waveguide distance of 2 μm. While coupling efficiency is limited around 20% in the SWW, due to the bow-shaped bending, in the R-SOLNET, the coupling efficiency reaches around 90% by the effect of the luminescent target.

4.7.2 MISALIGNMENT TOLERANCE IN PARALLEL R-SOLNETS

Figure 4.74 shows formation of parallel R-SOLNETs with luminescent targets for a waveguide distance of 4 μm. When lateral misalignments are 0 and 600 nm, by introducing 400-nm write beams from the input waveguides, parallel waveguides of R-SOLNET are pulled to the luminescent target locations, and 650-nm probe beams are guided toward them. For a lateral misalignment of 900 nm, however, R-SOLNETs toward the luminescent targets are not formed, accordingly, the probe beam goes straight without the self-alignment effect.

When 650-nm write beams are used instead of the 400-nm write beams as shown in Figure 4.75, for lateral misalignments of 0 and 600 nm, parallel waveguides of R-SOLNET are pulled to the luminescent target locations, and 800-nm probe beams are guided toward them. For a lateral misalignment of 1200 nm, parallel waveguides

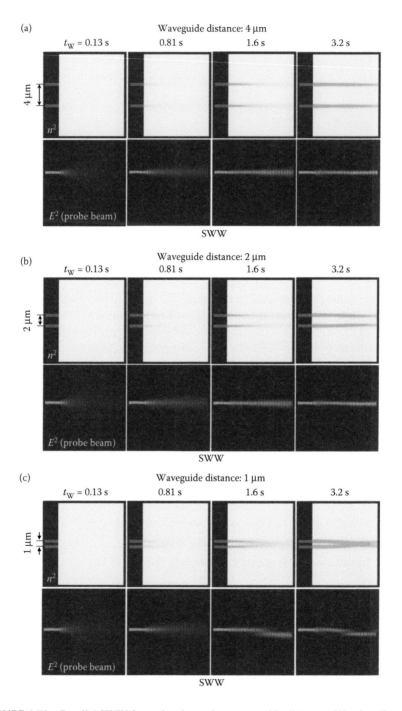

FIGURE 4.70 Parallel SWW formation for various waveguide distances. Wavelengths of a write beam, luminescence, and a probe beam are respectively 400, 500, and 650 nm. (From T. Yoshimura and M. Seki, *J. Opt. Soc. Am. B* **30**, 1643–1650, 2013.) [10]

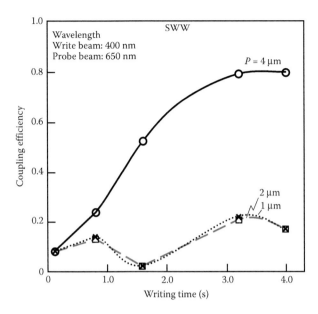

FIGURE 4.71 Dependence of coupling efficiency on writing time in parallel SWWs formed by 400-nm write beams for various waveguide distances. (From T. Yoshimura and M. Seki, *J. Opt. Soc. Am. B* **30**, 1643–1650, 2013.) [10]

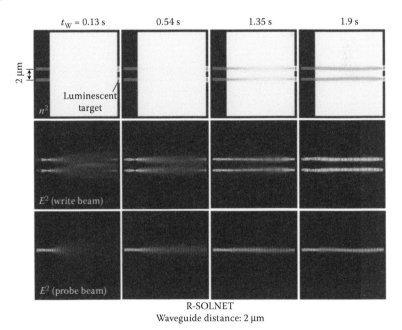

FIGURE 4.72 Parallel R-SOLNET formation for a waveguide distance of 2 μm. Wavelengths of a write beam, luminescence, and a probe beam are respectively 400, 500, and 650 nm. (From T. Yoshimura and M. Seki, *J. Opt. Soc. Am. B* **30**, 1643–1650, 2013.) [10]

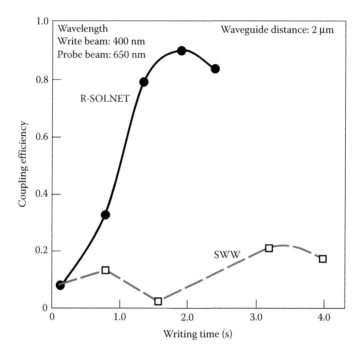

FIGURE 4.73 Dependence of coupling efficiency on writing time in parallel R-SOLNETs and SWWs formed by 400-nm write beams for a waveguide distance of 2 μm. (From T. Yoshimura and M. Seki, *J. Opt. Soc. Am. B* **30**, 1643–1650, 2013.) [10]

are still pulled to the luminescent target locations although the probe beam partially leaks. For a lateral misalignment of 1800 nm, the R-SOLNETs are not formed, but, separated waveguides are grown from the input waveguides and the luminescent targets. Accordingly, the probe beam goes straight without the self-alignment effect.

The coupling efficiency of a probe beam to the luminescent target for the write-beam wavelength of 400 nm is shown in Figure 4.76a. For a lateral misalignment of 0 nm, the efficiency exceeds 90%. For a lateral misalignment of 600 nm, coupling efficiency still reaches over 80%. For a lateral misalignment of 900 nm, the coupling efficiency drastically decreases. When 650-nm write beams are used, as shown in Figure 4.76b, for lateral misalignments of 0 and 600 nm, coupling efficiency reaches around 80%, which is less than the efficiency for the case of the 400-nm write beams. For a lateral misalignment of 1200 nm, coupling efficiency reaches around 60%, which is larger than the efficiency for the case of the 400-nm write beams. For a lateral misalignment of 1800 nm, very low coupling efficiency is observed. In all the cases, the crosstalk was calculated to be less than −23 dB. Thus, it is found that parallel R-SOLNETs with luminescent targets are available in lateral misalignment ranges of 0–600 nm for the write-beam wavelength of 400 nm, and in lateral misalignment ranges of 0–1200 nm for the write-beam wavelength of 650 nm.

These results are consistent with the write-beam divergence in the PRI material. The spread angle θ of write beams emitted from the 600-nm-wide input waveguide in

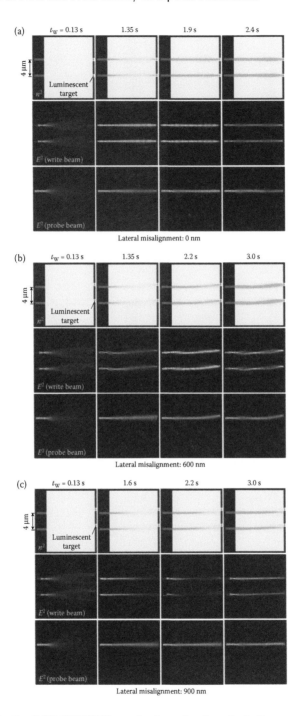

FIGURE 4.74 Parallel R-SOLNET formation for various lateral misalignments. Wavelengths of a write beam, luminescence, and a probe beam are respectively 400, 500, and 650 nm. (From T. Yoshimura and M. Seki, *J. Opt. Soc. Am. B* **30**, 1643–1650, 2013.) [10]

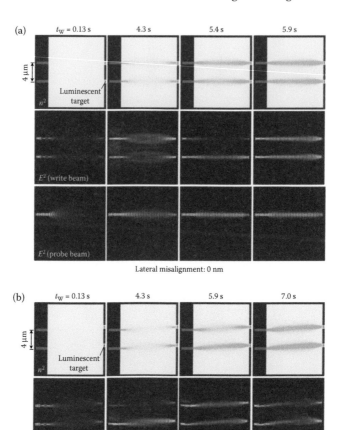

FIGURE 4.75   Parallel R-SOLNET formation for various lateral misalignments. Wavelengths of a write beam, luminescence, and a probe beam are respectively 650, 700, and 800 nm. (From T. Yoshimura and M. Seki, *J. Opt. Soc. Am. B* **30**, 1643–1650, 2013.) [10]   *(Continued)*

the PRI material with *n* of 1.5 are approximately estimated by Equation 4.7 to be 0.07 and 0.12 rad for write-beam wavelengths of 400 and 650 nm, respectively. The write-beam propagating distance from the input waveguide core edge to the luminescent target is 13 μm. From these parameters, it is found that write-beam illuminating portions extend to ∼920-nm wide from the input waveguide core axis for the 400-nm write beam, and ∼1500-nm wide for the 650-nm write beam. The wider illuminating portion gives rise to the wider lateral misalignment tolerance in the R-SOLNET formed by the 650-nm write beam.

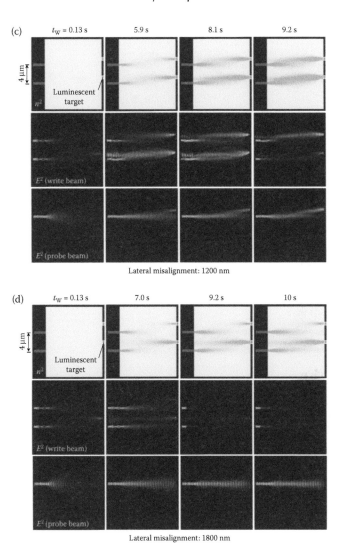

FIGURE 4.75 (Continued) Parallel R-SOLNET formation for various lateral misalignments. Wavelengths of a write beam, luminescence, and a probe beam are respectively 650, 700, and 800 nm. (From T. Yoshimura and M. Seki, *J. Opt. Soc. Am. B* **30**, 1643–1650, 2013.) [10]

It is found from Figures 4.74 and 4.75 that the width of R-SOLNETs formed by 650-nm write beams is wider than the width of those formed by 400-nm write beams, making the maximum coupling efficiency smaller for 650 nm than for 400 nm (see Figure 4.76). This can also be attributed to the larger divergence angle of 650-nm write beams compared to that of 400-nm write beams.

Although the simulations were performed only for the one-photon parallel R-SOLNETs, the similar effect of self-alignments is expected in the two-photon parallel R-SOLNETs.

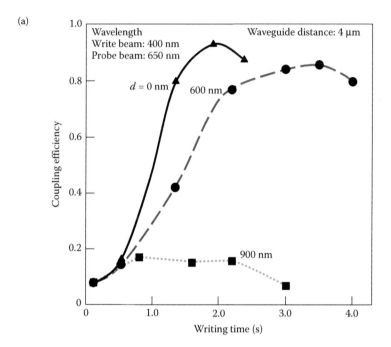

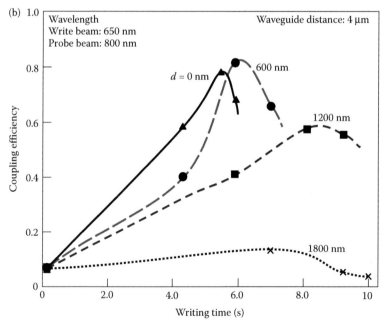

FIGURE 4.76 Dependence of coupling efficiency on writing time in parallel R-SOLNETs formed by (a) 400-nm write beams and (b) 650-nm write beams for various lateral misalignments. (From T. Yoshimura and M. Seki, *J. Opt. Soc. Am. B* **30**, 1643–1650, 2013.) [10]

4.8   Y-BRANCHING SOLNETs

SOLNETs with branching structures might expand their application fields, including cancer therapy. In this section, formation of Y-branching TB-SOLNETs and R-SOLNETs is demonstrated by the 2D FDTD method.

4.8.1   Y-Branching TB-SOLNET

A model for 2D FDTD simulations of Y-branching TB-SOLNETs is shown in Figure 4.77. The parameters in the model are the same as those described in Section 4.2.1 for Figure 4.1a, except that the number of output waveguides is two in the present model. d represents a lateral misalignment between the center axis of the input waveguide and the axis located at the center of the two output waveguides. The branching distance is denoted by b.

Formation of one-photon Y-branching TB-SOLNETs with a branching distance of 1200 nm is presented in Figure 4.78. For no lateral misalignments, by introducing a 500-nm write beam from the input waveguide and 600-nm write beams from the two output waveguides, self-focusing of the write beams is developed with writing time, resulting in a one-photon Y-branching TB-SOLNET between the input waveguide and the two output waveguides. At 2.7 s, due to overexposure, the branches merge to become one tapered waveguide. For a lateral misalignment of 600 nm, the Y-branching SOLNET is still formed. In two-photon Y-branching TB-SOLNETs, as shown in Figure 4.79, Y-branching structures are similarly formed. For a lateral misalignment of 600 nm due to a zigzag structure caused by overexposure, considerable probe-beam leakage is observed.

When the branching distance increases to 2400 nm, in the one-photon TB-SOLNET, in addition to the Y-branching structure, a straight waveguide is simultaneously formed by a straight-propagating write beam, resulting in a structure with three branches as can be seen in Figure 4.80a. In the case of a 600-nm lateral misalignment, the additional straight waveguide is trapped by the lower branch with considerable broadening as shown in Figure 4.80b. In two-photon TB-SOLNETs, Y-branching structures with less broadening are obtained both for 0-nm and 600-nm lateral misalignments as shown in Figure 4.81.

When the branching distance further increases to 3600 nm, as shown in Figure 4.82, the one-photon TB-SOLNET has a three-branch structure as in the case of the 2400-nm branching distance. In the two-photon TB-SOLNETs, as shown in

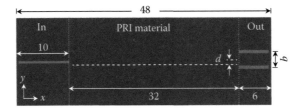

FIGURE 4.77   Model for 2D FDTD simulations of Y-branching TB-SOLNETs between a 600-nm-wide input waveguide and two 600-nm-wide output waveguides (units: μm).

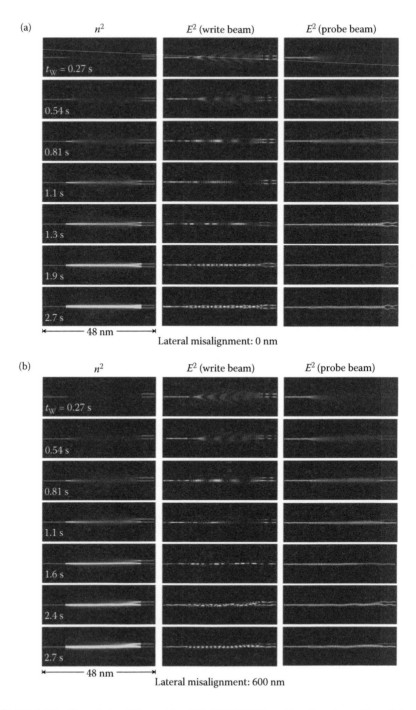

FIGURE 4.78 One-photon Y-branching TB-SOLNET formation for a branching distance of 1200 nm.

(a)

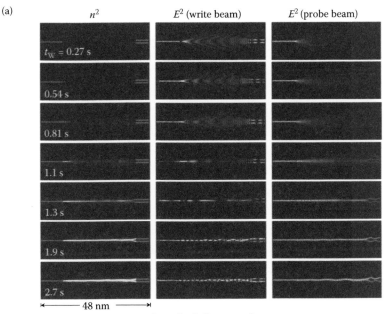

Lateral misalignment: 0 nm

(b)

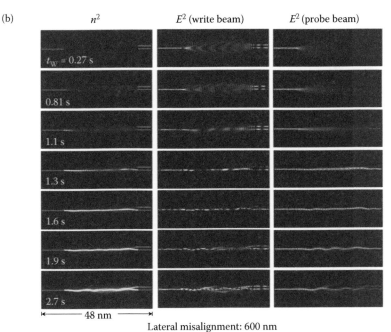

Lateral misalignment: 600 nm

FIGURE 4.79 Two-photon Y-branching TB-SOLNET formation for a branching distance of 1200 nm.

(a)

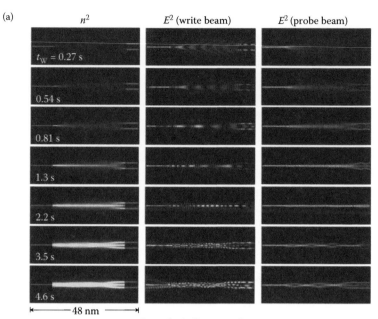

Lateral misalignment: 0 nm

(b)

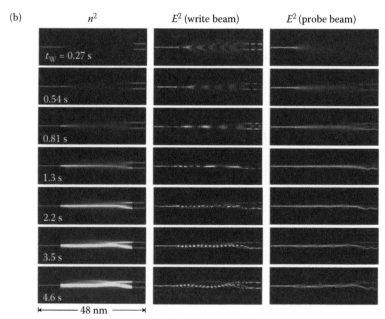

Lateral misalignment: 600 nm

FIGURE 4.80 One-photon Y-branching TB-SOLNET formation for a branching distance of 2400 nm.

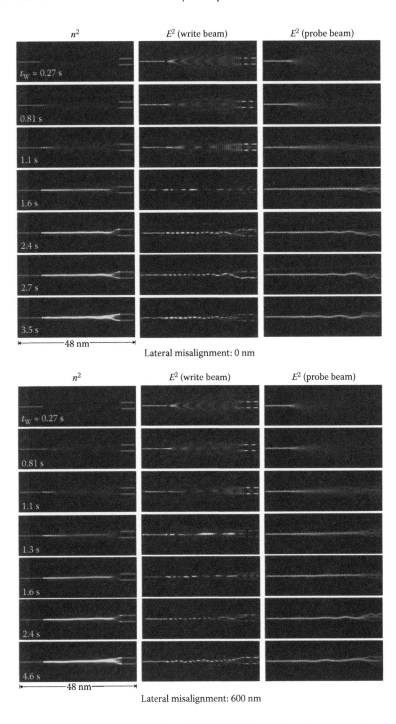

FIGURE 4.81   Two-photon Y-branching TB-SOLNET formation for a branching distance of 2400 nm.

n^2 E^2 (write beam) E^2 (probe beam)

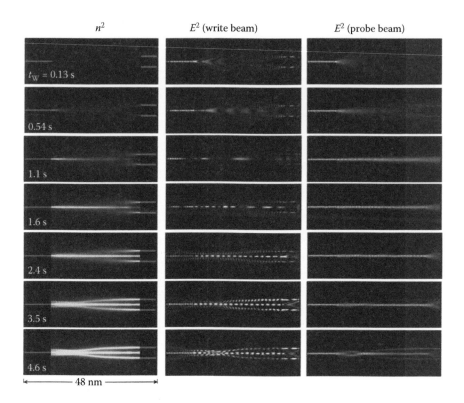

← 48 nm →

FIGURE 4.82 One-photon Y-branching TB-SOLNET formation for a branching distance of 3600 nm with no misalignments.

Figure 4.83, n^2, E^2 (write), and E^2 (probe) are symmetrically distributed about the center axis of the input waveguide until writing time of 1.1 s. After 1.3 s, a waveguide begins to stretch toward the upper output waveguide, and the write beam and probe beam are guided toward the upper output waveguide to complete an optical coupling between the input waveguide and the upper output waveguide at 2.2 s. At 3.5 s a write beam leaking from the coupling waveguide is observed, which forms another coupling waveguide to the lower output waveguide at 4.6 s. Thus, a two-photon Y-branching TB-SOLNET seems to be formed between the input waveguide and the two output waveguides. However, as can be seen in the probe-beam distribution at 4.6 s, the probe beam is not coupled to the output waveguides, indicating that the two-photon TB-SOLNET does not work as an effective Y-branching waveguide.

To form Y-branching SOLNETs with wider branching distances, some improvements in the write-beam condition, for example, an increase in the spread angle of the write beam from the input waveguide by waveguide lenses, might be necessary.

It is curious that in Figure 4.83 a coupling waveguide stretches toward one of the two output waveguides despite that the model is symmetry about the center axis of the input waveguide. This result might be attributed to some slight asymmetric modeling

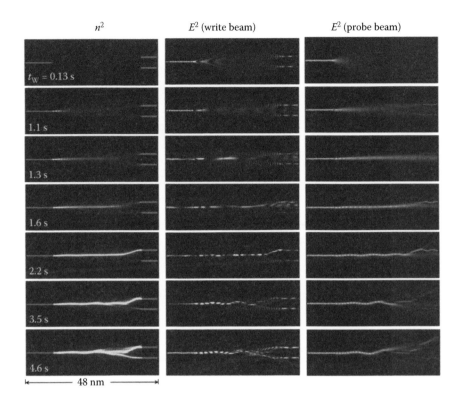

n^2 E^2 (write beam) E^2 (probe beam)

48 nm

FIGURE 4.83 Two-photon Y-branching TB-SOLNET formation for a branching distance of 3600 nm with no misalignments.

factors, which are amplified during the calculations. In other words, the SOLNET has steeply-selecting characteristics. Namely, once a trigger to pull the SOLNET, which is stronger than triggers from the other waveguides, is generated from one of the waveguides, a coupling waveguide tends to be selectively self-organized toward the waveguide.

4.8.2   Y-Branching R-SOLNET

A model for one-photon Y-branching R-SOLNETs with wavelength filters is shown in Figure 4.84a [19]. The widths of the input waveguide and two output waveguides are respectively 2 μm and 500 nm. The branching distance is 1000 nm. The refractive indices of the core and cladding region are 1.5 and 1.0, respectively. Refractive index of the PRI material changes from 1.5 to 1.8 upon write-beam exposure. Dielectric multilayer wavelength filters on the front edges of the output waveguides reflect 650-nm write beams. The filters consist of 80-nm-thick layers with a refractive index of 1.5 and 80-nm-thick layers with a refractive index of 2.5. The total layer count is four.

It can be seen from Figure 4.84b that a Y-branching structure is formed by the "pulling water" effect.

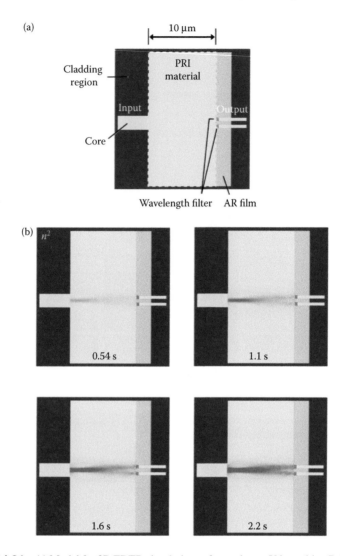

FIGURE 4.84 (a) Model for 2D FDTD simulations of one-photon Y-branching R-SOLNETs with wavelength filters between a 2-μm-wide waveguide and two 500-nm-wide waveguides. (b) One-photon Y-branching R-SOLNET formation with wavelength filters.

REFERENCES

1. T. Yoshimura, W. Sotoyama, K. Motoyoshi, T. Ishitsuka, K. Tsukamoto, S. Tatsuura, H. Soda, and T. Yamamoto, "Method of Producing Optical Waveguide System, Optical Device and Optical Coupler Employing The Same, Optical Network and Optical Circuit Board," U.S. Patent 6,081,632, 2000.
2. T. Yoshimura, J. Roman, Y. Takahashi, W. V. Wang, M. Inao, T. Ishitsuka, K. Tsukamoto, K. Motoyoshi, and W. Sotoyama, "Self-Organizing Waveguide Coupling Method 'SOLNET' and its Application to Film Optical Circuit Substrates," *Proc. 50th Electronic Components & Technology Conference (ECTC)*, Las Vegas, Nevada, 962–969, 2000.

3. T. Yoshimura, J. Roman, Y. Takahashi, W. V. Wang, M. Inao, T. Ishituka, K. Tsukamoto, K. Motoyoshi, and W. Sotoyama, "Self-Organizing Lightwave Network (SOLNET) and Its Application to Film Optical Circuit Substrates," *IEEE Trans. Comp., Packag. Technol.* **24**, 500–509, 2001.

4. T. Yoshimura, M. Ojima, Y. Arai, and K. Asama, "Three-Dimensional Self-Organized Micro Optoelectronic Systems for Board-Level Reconfigurable Optical Interconnects— Performance Modeling and Simulation—," *IEEE J. Select. Topics in Quantum Electron.* **9**, 492–511, 2003.

5. T. Yoshimura, T. Inoguchi, T. Yamamoto, S. Moriya, Y. Teramoto, Y. Arai, T. Namiki, and K. Asama, "Self-Organized Lightwave Network Based on Waveguide Films for Three-Dimensional Optical Wiring Within Boxes, *J. Lightwave Technol.* **22**, 2091–2100, 2004.

6. T. Yoshimura and H. Nawata, "Micro/Nanoscale Self-Aligned Optical Couplings of the Self-Organized Lightwave Network (SOLNET) Formed by Excitation Lights from Outside," *Opt. Commun.* **383**, 119–131, 2017.

7. S. Ono, T. Yoshimura, T. Sato, and J. Oshima, "Fabrication and Evaluation of Nano-Scale Optical Circuits Using Sol-Gel Materials with Photo-Induced Refractive Index Variation Characteristics," *J. Lightwave Technol.* **27**, 1229–1235, 2009.

8. F. E. Doany, B. G. Lee, S. N. Assefa, W. M. J. Green, M. Yang, C. L. Schow, C. V. Jahnes et al. "Multichannel High-Bandwidth Coupling of Ultradense Silicon Photonic Waveguide Array to Standard-Pitch Fiber Array," *J. Lightwave Technol.* **29**, 475–482, 2011.

9. S. Ono, T. Yoshimura, T. Sato, and J. Oshima, "Fabrication of Self-Organized Optical Waveguides in Photo-Induced Refractive Index Variation Sol-Gel Materials with Large Index Contrast," *J. Lightwave Technol.* **27**, 5308–5313, 2009.

10. T. Yoshimura and M. Seki, "Simulation of Self-Organized Parallel Waveguides Targeting Nanoscale Luminescent Objects," *J. Opt. Soc. Am. B* **30**, 1643–1650, 2013.

11. T. Yoshimura, "Simulation of Self-Aligned Optical Coupling between Micro- and Nanoscale Devices using Self-Organized Waveguides," *J. Lightwave Technol.* **33**, 849–856, 2015.

12. C. Brauchle, U. P. Wild, D. M. Burland, G. C. Bjorkund, and D. C. Alvares, "Two-Photon Holographic Recording with Continuous-Wave Lasers in the 750–1100-nm Range," *Opt. Lett.* **7**, 177–179, 1982.

13. T. Yoshimura, A. Hori, Y. Yoshida, Y. Arai, H. Kurokawa, T. Namiki, and K. Asama, "Coupling Efficiencies in Reflective Self-Organized Lightwave Network (R-SOLNET) Simulated by the Beam Propagation Method," *IEEE Photon. Technol. Lett.* **17**, 1653–1655, 2005.

14. T. Yoshimura, M. Iida, and H. Nawata, "Self-Aligned Optical Couplings by Self-Organized Waveguides toward Luminescent Targets in Organic/Inorganic Hybrid Materials," *Opt. Lett.* **39**, 3496–3499, 2014.

15. K. Takahashi, "Research and Development on ultra-high-density 3-dimensional LSI-chip-stack packaging technologies," *The 3rd Annual Meeting on Electronics System Integration Technologies Digest* (edited and published by the Electronic System Integration Technology Research Department, Association of Super-Advanced Electronics Technologies (ASET)), 43–94, 2002.

16. T. Yoshimura, D. Takeda, T. Sato, Y. Kinugasa, and H. Nawata, "Polymer Waveguides Self-Organized by Two-Photon Photochemistry for Self-Aligned Optical Couplings with Wide Misalignment Tolerances," *Opt. Commun.* **362**, 81–86, 2016.

17. H. Kaburagi and T. Yoshimura, "Simulation of Micro/Nanometer-Scale Self-Organized Lightwave Network (SOLNET) Using the Finite Difference Time Domain Method," *Opt. Commun.* **281**, 4019–4022, 2008.

18. T. Yoshimura, K. Wakabayashi, and S. Ono, "Analysis of Reflective Self-Organized Lightwave Network (R-SOLNET) for Z-Connections in Three-Dimensional Optical Circuits by the Finite Difference Time Domain Method," *IEEE J. Select. Topics in Quantum Electron.* **17**, 566–570, 2011.
19. T. Yoshimura and H. Kaburagi, "Self-Organization of Optical Waveguides between Misaligned Devices Induced by Write-Beam Reflection," *Appl. Phys. Express* **1**, 062007, 2008.

5 Preferable Waveguide Growth Condition for SOLNET Formation

In this chapter, after showing examples of straight and vertical waveguide growth in PRI materials, influence of write-beam absorption and intensity in the PRI materials on the waveguide growth are described in order to clarify preferable conditions for SOLNET formation.

For experiments using photopolymers or SUNCONNECT®, two kinds of optical fibers are used: a 50-μm optical fiber (an optical fiber with a core diameter of ~50 μm), which is a multimode fiber for 1.3 μm in wavelength, and a 9-μm optical fiber, which is a single-mode fiber for 1.3 μm. For experiments using PRI sol-gel materials, a 3-μm optical fiber, which is a single-mode fiber for 405 nm, is used.

5.1 STRAIGHT WAVEGUIDE GROWTH

An interference microscopic photograph shown in Figure 5.1a is an example of growth of a straight optical waveguide of SWW at writing time of 120 s [1]. A 80-nW UV write beam of a high-pressure mercury lamp was introduced from a core of a 9-μm optical fiber into a free space filled with a monomer/binder-type photopolymer, which consists of carbazole monomers for high-refractive-index components and a binder of acrylic (or epoxy) polymers for a low-refractive-index host matrix with addition of sensitizing agents and polymerization initiator. The maximum refractive index change induced by a write beam in the photopolymer is typically ~0.02. A straight line stretching from the fiber core along the write-beam propagation axis is the constructed straight optical waveguide, where the refractive index is higher than the surrounding area. The optical waveguide has a width of almost the same as the fiber core diameter. It can be seen from Figure 5.1b that the write beam spreads due to diffraction before optical waveguide growth, and after optical waveguide growth the write beam is confined in it.

Figure 5.2 shows direct optical waveguide growth from an LD [2]. A dye-sensitized monomer/binder-type photopolymer for red write beams was placed in front of a 680-nm wavelength LD. By emitting a write beam from the LD, an optical waveguide stretching from the LD appeared in the photopolymer in ~10 s, as can be seen in Figure 5.2a. The write beam from the LD was spontaneous emission with power of ~20 μW, and was modulated with a duty ratio of 1/10, implying that the effective write-beam power is ~2 μW. The LD light beam was confined into the optical waveguide as shown in Figure 5.2b. Coupling loss was measured between the LD and a 9-μm optical fiber placed in a counter configuration with no lateral

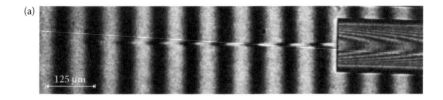

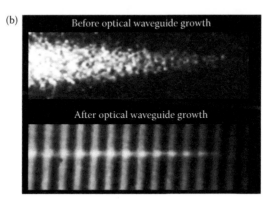

FIGURE 5.1   (a) Growth of a straight optical waveguide of SWW from a 9-μm optical fiber at writing time of 120 s. (b) Probe-beam propagation before and after the optical waveguide growth. (From T. Yoshimura et al., "Self-Organizing Waveguide Coupling Method "SOLNET" and its Application to Film Optical Circuit Substrates," *Proc. 50th Electronic Components & Technology Conference (ECTC)*, 962–969, 2000.) [1]

misalignments and a 100-μm gap distance. The waveguide growth decreased the coupling loss from 14 to 7 dB.

In optical waveguides grown from LDs directly, the wavelengths of write beams and signal beams are the same. This causes a concern of an insertion loss increase due to signal beam absorption by photopolymers. However, the sensitizing dyes are bleached by absorbing write beams to reduce the signal beam absorption after the waveguide

FIGURE 5.2   (a) Growth of an optical waveguide from 680-nm LD and (b) LD light-beam propagation in the optical waveguide. (From T. Yoshimura et al. *J. Lightwave Technol.* **22**, 2091–2100, 2004.) [2]

growth because the photochemical reactions are initiated by dissociation of the dyes. Furthermore, because the length of the grown optical waveguides is typically shorter than several hundred µm, signal beam absorption by the photopolymers does not affect largely on insertion losses.

Another concern is post curing of the photopolymer. After growing an optical waveguide by a write beam, a signal beam with the same wavelength as that of the write beam is transmitted in the optical waveguide with much higher intensity. This causes further photochemical reactions in the photopolymer, destroying the refractive index distribution of the optical waveguide. To suppress the influence of the signal beam, post curing condition should be optimized for blanket UV light exposure or baking.

5.2   VERTICAL WAVEGUIDE GROWTH

A fabrication/measurement setup for vertical waveguides is shown in Figure 5.3 [3]. On the back of an under-cladding layer in an optical waveguide film, a PRI material layer of ~500-µm thickness is coated. For the PRI material, a monomer/monomer-type photopolymer developed by JSR Corporation [4] is used. It consists of acrylic component and epoxy component. The acrylic component contains an aromatic diacrylate with high-refractive index and a radical photoinitiator. The epoxy component contains an aliphatic diepoxide and a photoacid generator. The refractive index is 1.53 and 1.51 for the acrylic and epoxy components, respectively. The mixing ratio of the two components is 50:50 in wt%. The acrylic component exhibits larger reactivity than the epoxy component upon irradiation of light in the blue-violet regions, inducing the PRI phenomena. The photopolymer gives rise to a maximum refractive index change around 0.02.

A 405-nm write beam with power of ~100 µW or less is introduced from a 9-µm optical fiber into *Input* of the optical waveguide film. The propagated write beam is reflected to the surface-normal direction by a vertical mirror at the edge of the optical waveguide. The write beam passes through the under-cladding layer and enters the

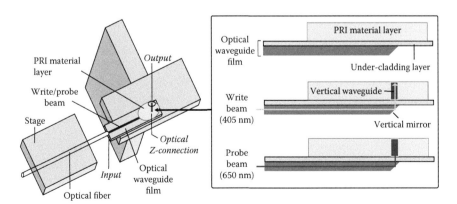

FIGURE 5.3   Experimental setup for growth of vertical waveguides. (From T. Yoshimura et al., *J. Lightwave Technol.* **24**, 4345–4352, 2006.) [3]

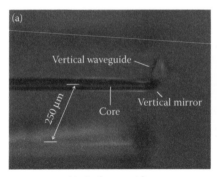
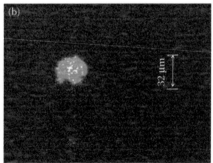

Vertical waveguide   Near field pattern
(View from above with an angle of ~30°)

FIGURE 5.4   (a) Vertical waveguide grown above the vertical mirror of an optical waveguide film. (b) An NFP of the probe beam guided in the vertical waveguide at the PRI material layer surface. (From T. Yoshimura et al. *J. Lightwave Technol.* **24**, 4345–4352, 2006.) [3]

PRI material layer to grow a vertical waveguide. Next, a 650-nm probe beam is introduced into the *Input*. The beam is guided in the vertical waveguide at *Optical Z-Connection* to reach *Output* at the PRI material layer surface.

Figure 5.4a shows a vertical waveguide grown in the PRI material layer above the vertical mirror aperture. As shown in Figure 5.4b, at *Output*, a near field pattern (NFP) of the guided probe beam is observed. The NFP size is $\sim 30 \times 30$ μm^2, which is approximately the same as the core size of the optical waveguide film. These results demonstrate an optical via capability of the vertical waveguide for 3D optical wiring.

It should be noted that the vertical waveguide is grown by a write beam from the under-cladding layer, which is a free space with no beam-confining structures. In other words, the vertical waveguide can be grown even when some spacers are inserted between vertical mirrors and PRI material layers.

5.3   INFLUENCE OF WRITE-BEAM ABSORPTION AND INTENSITY ON WAVEGUIDE GROWTH

Figure 5.5 schematically shows how optical waveguide shapes are influenced by write-beam absorption and write-beam intensity in PRI materials. When the write-beam absorption is too large, the write beam propagates only a short distance, growing a short waveguide. Once a waveguide is grown, the absorption in the waveguide region decreases due to decomposition of sensitizers, and then, next-step waveguide growth starts to grow another short waveguide. By repeating this process, a whole waveguide is constructed step-by-step, just like "a mole digs a tunnel to go forward." This limits the write beam spreading required to increase the write-beam overlapping in SOLNET formation processes.

When the write-beam absorption is small enough, on the other hand, a write beam can propagate over a long distance to induce uniform self-focusing all over the PRI materials, growing a long uniform optical waveguide. This allows us to increase the write-beam overlapping, which is favorable for SOLNET formation.

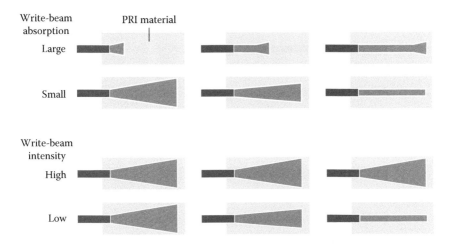

FIGURE 5.5 Influence of the write-beam absorption and the write-beam intensity in PRI materials on grown waveguide shapes.

For write-beam intensity, when the intensity is too high, molecular reactions reach a saturated condition before the molecular diffusion occurs. This prevents self-focusing. In such cases, the constructed optical waveguides tend to become spreading and scattered. So, the write-beam intensity should be low enough to promote smooth self-focusing.

5.3.1 WRITE-BEAM ABSORPTION

In order to examine the influence of the write-beam absorption on the optical waveguide growth dynamics, optical waveguides were grown in dye-sensitized photopolymers. Dye sensitization is effective to control not only the spectral responses but also the amount of the write-beam absorption. The photopolymers were prepared by mixing the acrylic materials of NOA81 (NORLAND) with refractive index of 1.56 and NOA65 with refractive index of 1.52. The mixing weight ratio was NOA81:NOA65 = 2:1. Crystal violet (CV) was added into the mixture as a sensitizer. Optical waveguides were grown by emitting 532-nm write beams from a 50-µm optical fiber into the photopolymers.

As Figure 5.6 shows, in a photopolymer containing 3 wt% CV, step-like short-distance optical waveguide growth is observed. In a photopolymer containing 0.15 wt% CV, long-distance uniform growth is observed as expected. Here, the bending of the waveguide may occur by slight flow of the photopolymer.

In Figure 5.6, although the optical waveguide growth was carried out by a green write beam of 532 nm in wavelength, the optical waveguide is observed as red images. This phenomenon is attributed to the red photoluminescence (PL) from CV doped in the photopolymer. The 532-nm write beam that acts as an excitation light for CV is confined in the grown optical waveguide to increase the excitation light intensity in the waveguide. Thus, the red images corresponding to the waveguide shape appear. This unique property is useful for monitoring the optical waveguide growth dynamics.

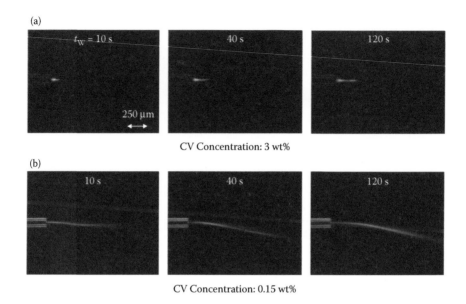

CV Concentration: 3 wt%

CV Concentration: 0.15 wt%

FIGURE 5.6 Influence of CV concentration in PRI materials on optical waveguide growth dynamics, which is monitored by red PL from CV excided by a 532-nm write beam.

Figure 5.7 shows another example to reveal the influence of the write-beam absorption on the waveguide growth dynamics. Optical waveguides were grown from 50-μm optical fibers in a PRI material of a photosensitive organic/inorganic hybrid material, SUNCONNECT® developed by Nissan Chemical Industries, Ltd

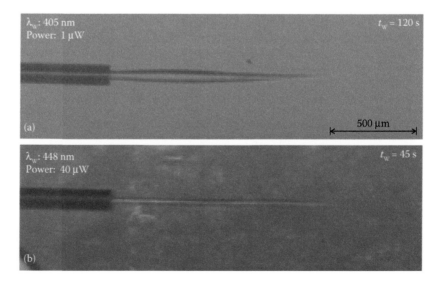

FIGURE 5.7 Optical waveguide growth from 50-μm fibers in SUNCONNECT® by write beams with wavelengths of (a) 405 nm and (b) 448 nm.

[5,6]. The refractive index contrast between the core and cladding region is expected to be ~1.60/1.59. The write-beam power required to grow a 1.5-mm-long optical waveguide is almost two orders of magnitude larger for the write-beam wavelength λw of 448 nm than for λw of 405 nm. This is attributed to the fact that the absorption coefficient of SUNCONNECT® is smaller at 448 nm than at 405 nm (see Figure 6.16 in Section 6.2.4.1).

As observed in Figure 5.7a, a nonuniform widened waveguide is grown when a 405-nm write beam is used. When a 448-nm write beam is used, as observed in Figure 5.7b, a straight waveguide is grown with a uniform width that is close to the fiber core width. This difference can be explained in terms of the differing write-beam absorption in the PRI material as indicated in Figure 5.5. Namely, if there is large write-beam absorption, a waveguide is grown in a stepwise manner. This might result in nonuniformity of the waveguide shape. Conversely, for small write-beam absorption, the write beam can propagate over a longer distance allowing uniform self-focusing to occur. This results in the growth of a straight waveguide with a more uniform shape.

From these results, it is confirmed that PRI materials with small write-beam absorption are suitable to grow uniformly-self-focused optical waveguides. This leads us to believe that the small-absorption-limit condition is preferable for the formation of SOLNETs because the write-beam-overlapping area is widened in such condition.

5.3.2 WRITE-BEAM INTENSITY

The influence of the write-beam intensity on the optical waveguide growth dynamics were examined by using the PRI sol-gel material developed by Nissan Chemical Industries, Ltd. [7], whose detail is described in Section 6.2.3. The material exhibits a large refractive index increase from 1.65 to 1.85 by UV/blue-light exposure under heating. The increase rate of the refractive-index increases with the heating temperature. This implies that "growth of optical waveguides at fixed write-beam intensity with increased heating temperature" is equivalent to "growth of optical waveguides at a fixed heating temperature with increased write-beam intensity."

Figure 5.8 shows optical waveguides grown in a PRI sol-gel thin film, which is deposited on a Si wafer with a SiO_2 cladding layer, by introducing 405-nm write beams with power of 0.5 mW from a 3-μm optical fiber at various heating temperatures. The writing time is 60 s. As described above, the results in Figure 5.8 can be regarded that the write-beam intensity increases in the order of (a), (b), (c), and (d), representing the influence of the write-beam intensity on the optical waveguide growth dynamics.

It is found from Figure 5.8 that straight optical waveguide growth is observed when the heating temperature is 100°C. By increasing the temperature to 150°C and 200°C, by increasing the write-beam intensity equivalently, the grown optical waveguide shapes become spreading and scattering as expected (see Figure 5.5). At a heating temperature of 50°C, little trace of optical waveguide growth is observed, suggesting that too small write-beam intensity is not a suitable condition.

Photographs of write-beam propagation before and after optical waveguide growth are shown in Figure 5.9 for write-beam power of 0.1, 0.5, and 1.0 mW. In each case,

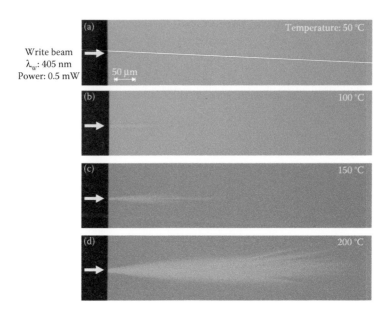

FIGURE 5.8 Optical waveguide growth in PRI sol-gel thin films at various heating temperatures. The writing time is 60 s. (From S. Ono et al. *J. Lightwave Technol.* **27**, 5308–5313, 2009.) [7]

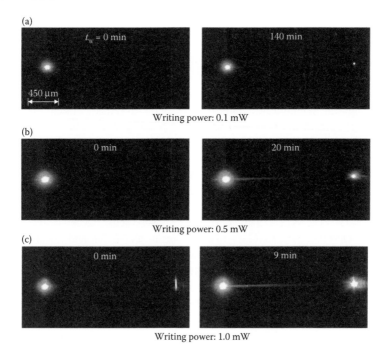

FIGURE 5.9 Write-beam propagation before and after optical waveguide growth for various writing power. (From S. Ono et al. *J. Lightwave Technol.* **27**, 5308–5313, 2009.) [7]

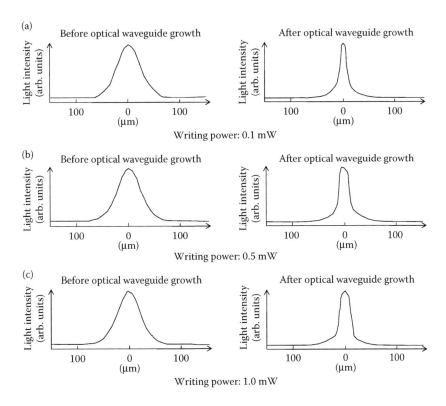

FIGURE 5.10 NFPs of output write beams of 405 nm in wavelength from the output edge of the PRI sol-gel thin film before and after optical waveguide growth for various writing power. (From S. Ono et al. *J. Lightwave Technol.* **27**, 5308–5313, 2009.) [7]

as writing time advances, an optical waveguide is stretched and reaches the right-side edge, where, a confined beam spot is observed.

The NFPs of the outputs of write beams are shown in Figure 5.10. In all of the three cases, the output write-beam spot width becomes narrow after optical waveguide growth. The relationship between the write-beam power and the half-width of the output beam spot, which is a measure of the write-beam confinement, is shown in Figure 5.11 The half-width after optical waveguide growth decreases from 24 to 12 μm as write-beam power decreases from 1.0 to 0.1 mW. The result indicates that the widths of the grown optical waveguides become narrow as write-beam power is reduced, suggesting that optical waveguides grown with low write-beam power exhibit strong self-focusing characteristics. By reducing the write beam power further, narrowing of the optical waveguide to a submicron scale is expected as predicted by the FDTD simulations in Sections 4.2 and 4.3.

Also, the half-width of the output beam spot along the film thickness direction was estimated to be 300 nm, which is determined by the thickness of the PRI sol-gel thin film.

The influence of the write-beam power on the relative coupling efficiency is shown in Figure 5.12 for probe-beam wavelengths of 405 and 633 nm. The relative coupling

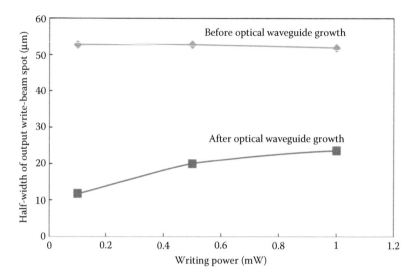

FIGURE 5.11   Influence of writing power on the write-beam confinement at 405 nm in wavelength. (From S. Ono et al. *J. Lightwave Technol.* **27**, 5308–5313, 2009.) [7]

efficiency is a ratio of B/A. Here, A and B are the light-beam power received by a 9-μm optical fiber at the output edge of the PRI sol-gel thin film, respectively, before and after optical waveguide growth. In each case, the coupling efficiency increases as write-beam power is reduced. The result confirms that optical waveguides grown with low write-beam power provide strong light-beam confinement effects.

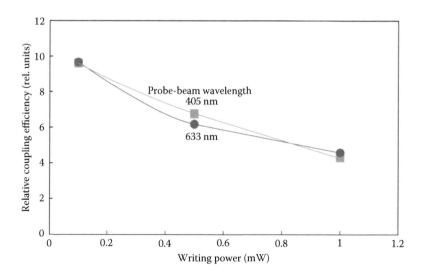

FIGURE 5.12   Influence of write-beam power on the relative coupling efficiency for probe beams of 405 nm and 633 nm in wavelength. (From S. Ono et al. *J. Lightwave Technol.* **27**, 5308–5313, 2009.) [7]

Thus, it is concluded from these results that a low write-beam-intensity condition is suitable to grow uniformly-self-focused optical waveguides, which is desirable for the formation of SOLNETs, because slow photochemical reactions enough to achieve gradual self-focusing can be performed.

REFERENCES

1. T. Yoshimura, J. Roman, Y. Takahashi, W. V. Wang, M. Inao, T. Ishitsuka, K. Tsukamoto, K. Motoyoshi, and W. Sotoyama, "Self-Organizing Waveguide Coupling Method "SOLNET" and its Application to Film Optical Circuit Substrates," *Proc. 50th Electronic Components & Technology Conference (ECTC)*, Las Vegas, Nevada, 962–969, 2000.
2. T. Yoshimura, T. Inoguchi, T. Yamamoto, S. Moriya, Y. Teramoto, Y. Arai, T. Namiki, and K. Asama, "Self-Organized Lightwave Network Based on Waveguide Films for Three-Dimensional Optical Wiring within Boxes," *J. Lightwave Technol.* **22**, 2091–2100, 2004.
3. T. Yoshimura, M. Miyazaki, Y. Miyamoto, N. Shimoda, A. Hori, and K. Asama, "Three-Dimensional Optical Circuits Consisting of Waveguide Films and Optical Z-Connections," *J. Lightwave Technol.* **24**, 4345–4352, 2006.
4. F. Huang, H. Takase, Y. Eriyama, and T. Ukachi, "Optical Fiber Interconnection by Using Self-Written Waveguides," *Proc. 9th Int. Conf. Rad. Tech. Asia*, Japan, 637–639, 2003.
5. T. Sato, "Novel Organic-Inorganic Hybrid Materials for Optical Interconnects," *Proc. SPIE* **7944**, 79440M, 2011.
6. H. Nawata, "Organic-Inorganic Hybrid Material for on-Board Optical Interconnection and its Applications in Optical Coupling," *IEEE CPMT Symposium*, Japan, 121–124, 2013.
7. S. Ono, T. Yoshimura, T. Sato, and J. Oshima, "Fabrication of Self-Organized Optical Waveguides in Photo-Induced Refractive Index Variation Sol-Gel Materials with Large Index Contrast," *J. Lightwave Technol.* **27**, 5308–5313, 2009.

6 Experimental Demonstrations of SOLNETs

In this chapter, experimental demonstrations using optical fibers are presented for TB-SOLNETs and R-SOLNETs. Demonstrations of Y-branching SOLNETs and two-photon SOLNETs are also given.

For experiments using photopolymers or SUNCONNECT®, a 50-μm optical fiber and a 9-μm optical fiber are used. For experiments using PRI sol-gel materials, a 3-μm optical fiber is used.

6.1 TB-SOLNETs

The interference microscopic photograph shown in Figure 6.1 is the firstly-achieved experimental demonstration of the SOLNET [1,2]. Two 9-μm optical fibers are put in a glass cylinder with a counter direction. The gap between the optical fibers is filled with the monomer/binder-type photopolymer that is described in Section 5.1. The gap distance is 400 μm. By emitting 40-nW UV write beams of a high-pressure mercury lamp from the two optical fibers into the photopolymer, a coupling optical waveguide of TB-SOLNET connecting the two optical fibers is gradually formed with writing time. In Figure 6.1a, a TB-SOLNET formed by 300-s writing is shown.

The coupling efficiency between the two optical fibers was measured using a 1.3-μm probe beam, to which the photopolymer is insensitive. As shown in Figure 6.1b, with writing time, the coupling efficiency increases up to 80%, then decreases slightly toward a saturated value. The writing time dependence of the coupling efficiency is similar to that predicted by the BPM simulation for TB-SOLNETs between 8-μm-wide optical waveguides (see Figure 4.5a).

The decrease in the coupling efficiency beyond the peak might be caused by overexposure in the following two ways. Excessive change in the refractive index of the photopolymer due to the overexposure of write beams (1) increases the beam confinement ability of the SOLNET to make the mode size in the SOLNET smaller than that in input/output waveguides, or (2) increases the SOLNET width to make the mode size larger than that of input/output waveguides. These mode size mismatches cause loss at the interface between the SOLNET and the input/output waveguides, giving rise to the coupling efficiency reduction. After a sufficient amount of write-beam exposure, the write beams are completely confined in the SOLNET core, and at the same time, the refractive index reaches its saturation limit. Then, no more SOLNET core shape changes are induced, resulting in the coupling efficiency saturation.

Figure 6.2 shows TB-SOLNETs connecting two 9-μm optical fibers, where the monomer/monomer-type photopolymer that is described in Section 5.2 is used [3].

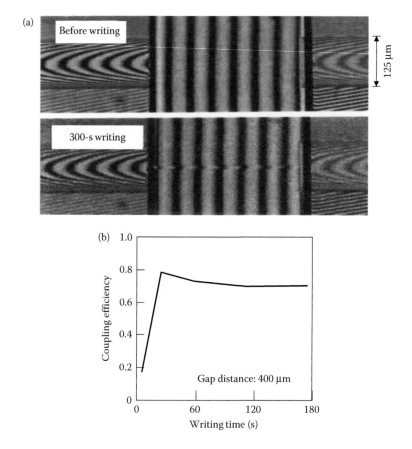

FIGURE 6.1 (a) TB-SOLNET formation between 9-μm optical fibers in the monomer/binder-type photopolymer. (b) Dependence of coupling efficiency on writing time in the TB-SOLNET. (From T. Yoshimura et al. "Self-Organizing Waveguide Coupling Method "SOLNET" and its Application to Film Optical Circuit Substrates," *Proc. 50th Electronic Components & Technology Conference (ECTC)*, 962–969, 2000.) [1]

For lateral misalignments of 1 μm–9 μm, self-aligned coupling waveguides of TB-SOLNET are formed. For a lateral misalignment of 20 μm, optical waveguides grown from the two optical fibers do not merge, but, remain as two separated optical waveguides. This is attributed to the insufficient overlap of the write beams emitted from the two optical fibers.

In Figure 6.3a, dependence of coupling efficiency on writing time in a TB-SOLNET is shown for a lateral misalignment of ∼1 μm. With writing time, the coupling efficiency increases up to 85% (coupling loss of 0.7 dB), then decreases slightly to reach a saturated level. The writing time dependence of the coupling efficiency is similar to that in the experiment using the monomer/binder-type photopolymer shown in Figure 6.1b, and also similar to that predicted by the BPM simulation (see Figure 4.5a). The results suggest that the prediction by the BPM simulation provides general characteristics of the TB-SOLNET formation independently of PRI materials.

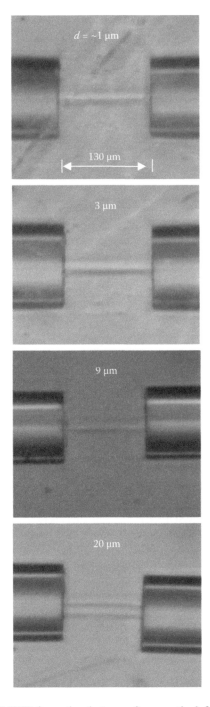

FIGURE 6.2   TB-SOLNET formation between 9-μm optical fibers for various lateral misalignments in the monomer/monomer-type photopolymer. Writing-beam intensity is 1.3 W/cm^2.

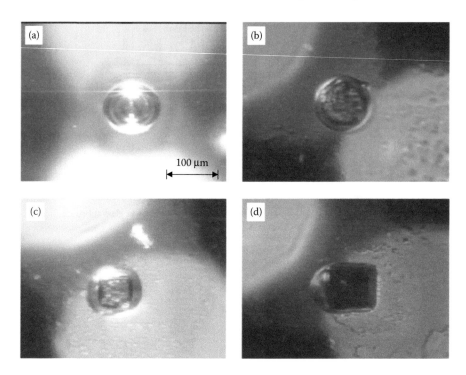

FIGURE 6.5 Photographs taken from the lower-substrate side by detaching the lower substrate after write-beam exposure. (a) An optical waveguide stretching from the circle window on the upper substrate, (b) R-SOLNET between the circle window and a circle mirror with no misalignments, (c) R-SOLNET between the circle window and a square mirror, which is smaller than the circle window, with no misalignments, and (d) R-SOLNET between the circle window and a square mirror with 30-μm misalignment. (From T. Yoshimura et al. *J. Lightwave Technol.* **22**, 2091–2100, 2004.) [4]

misalignment. A coupling optical waveguide with a circle bottom and a square top can be seen, where a displacement exists between the top and the bottom according to the misalignment between the circle window and the square mirror.

These results indicate that R-SOLNETs can be formed using free-space write beams.

6.2.2 R-SOLNET WITH MICROMIRRORS

R-SOLNET formation between a 50-μm optical fiber and an Al micromirror deposited on an optical fiber edge with angular misalignments is demonstrated using a setup shown in Figure 6.6a [5]. The 50-μm optical fiber and the micromirror are placed with a gap distance of ~800 μm in the monomer/monomer-type photopolymer. An angular misalignment of 3° exists between the fiber direction and the surface–normal direction of the micromirror.

Figure 6.6b shows a photograph of the 50-μm optical fiber and the optical fiber with the Al micromirror on the edge in the photopolymer before R-SOLNET

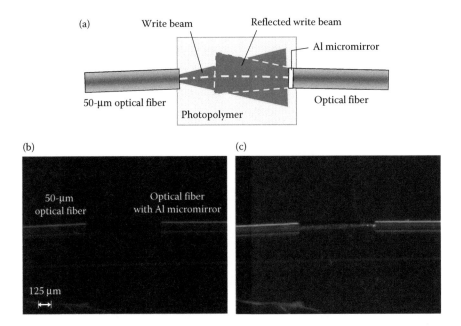

FIGURE 6.6   Formation of R-SOLNET between a 50-μm optical fiber and an Al micromirror on an optical fiber edge with an angular misalignment of 3°. (a) Experimental setup, (b) the 50-μm optical fiber and the optical fiber with an Al micromirror on the edge surface in a photopolymer before SOLNET formation, and (c) 650-nm probe-beam propagation in the bow-shaped self-aligned coupling waveguide of R-SOLNET.

formation. By introducing a UV write beam of a high-pressure mercury lamp from the 50-μm optical fiber into the photopolymer, an R-SOLNET was attempted to form. If the "pulling water" effect did not exist, two separated optical waveguides with a directional difference of 6° would be observed stretching from the micromirror: one would be grown by the incident write beam and the other by the reflected write beam. However, as Figure 6.6c shows, when the UV write beam is introduced, a bow-shaped R-SOLNET is formed between the optical fiber and the micromirror by the "pulling water" effect, and a 650-nm probe beam from the 50-μm optical fiber propagates in the R-SOLNET. The core width of the R-SOLNET is about 50 μm, which is close to that of the 50-μm optical fiber.

As shown in Figure 6.7, the bow-shaped R-SOLNETs can be formed when the angular misalignment is less than 6°. For angular misalignments of 10° and 40°, the "pulling water" effect vanishes. Consequently, R-SOLNETs cannot be formed. Instead, two separated optical waveguides stretching from the micromirror are observed.

R-SOLNET formation between a 50-μm optical fiber and an Al micromirror deposited on an optical fiber edge with lateral misalignments is demonstrated using a setup shown in Figure 6.8a [5]. Figure 6.8b shows a photograph before R-SOLNET formation. When a UV write beam is introduced from the 50-μm optical fiber into the photopolymer, an S-shaped R-SOLNET is formed, connecting

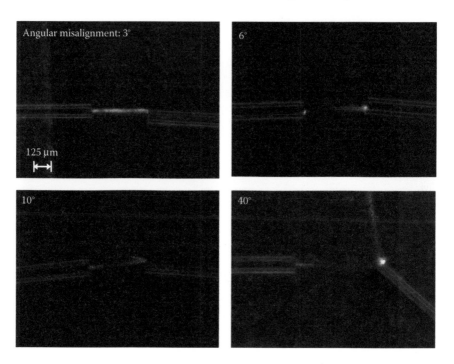

FIGURE 6.7 Formation of R-SOLNETs with micromirrors for various angular misalignments. (Collaboration with Takashi Kuma.)

the 50-μm optical fiber to the misaligned micromirror, and a 650-nm probe beam propagates in the S-shaped R-SOLNET with a core width of about 50 μm as shown in Figure 6.8c.

It is noted that the R-SOLNET is connected to the upper part of the fiber edge. As can be seen in Figure 6.9, the micromirror is deposited partially. This might be the reason why R-SOLNET is pulled to the upper part. Actually, a trace of the R-SOLNET is found on the upper part of the fiber edge, where the micromirror surface exists. Considering this situation, the actual lateral misalignment between the 50-μm optical fiber and the micromirror is estimated to be around 60 μm.

In order to measure the coupling efficiency during R-SOLNET formation, an R-SOLNET was formed by a semitransparent Al mirror. The setup for the experiments is shown in Figure 6.10a. An R-SOLNET is formed by an incident write beam from the 50-μm optical fiber and a reflected write beam from the semitransparent Al mirror deposited on a glass substrate. The write beam on the glass substrate surface is observed by a camera from the backside. The results are shown in Figure 6.10b. Initially, the write beam reaches the left side of the semitransparent mirror. With writing time, the write beam is pulled to the semitransparent mirror. Figure 6.11 shows coupling efficiency variation with writing time. The efficiency was monitored using a 50-μm optical fiber placed just behind the semitransparent Al mirror. With R-SOLNET formation, the coupling efficiency increases. After reaching the peak, the coupling efficiency decreases due to overexposure.

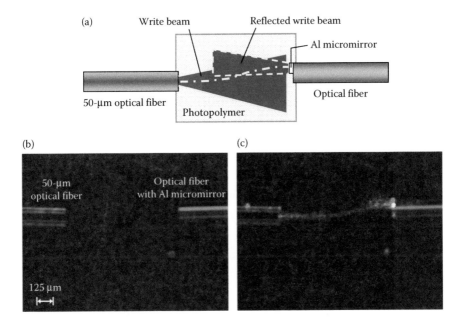

FIGURE 6.8 Formation of R-SOLNET between a 50-μm optical fiber and an Al micromirror on an optical fiber edge with a lateral misalignment of ~60 μm. (a) Experimental setup, (b) the 50-μm optical fiber and the optical fiber with an Al micromirror on the edge surface in a photopolymer before SOLNET formation, and (c) 650-nm probe-beam propagation in the S-shaped self-aligned coupling waveguide of R-SOLNET.

6.2.3 R-SOLNET with Reflective Objects

R-SOLNETs with reflective objects are demonstrated in the PRI sol-gel material developed by Nissan Chemical Industries, Ltd. [6,7]. It is a silicon-oxide-based material, which exhibits a large refractive index increase from ~1.65 to ~1.85 by UV/blue-light exposure and simultaneous heating, forming high-index-contrast (HIC) SOLNETs for nanoscale waveguides.

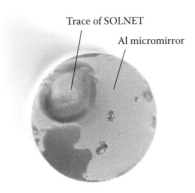

FIGURE 6.9 A photograph of the Al micromirror on the fiber edge surface.

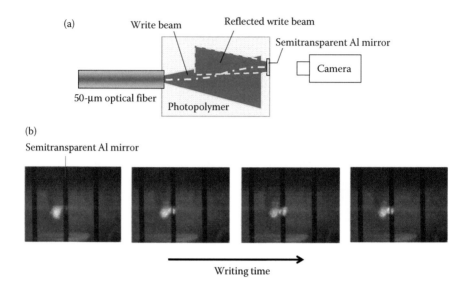

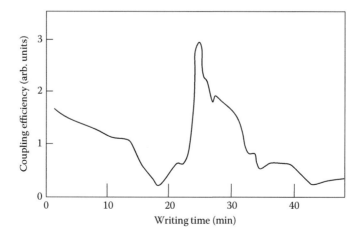

FIGURE 6.10 (a) Experimental setup for R-SOLNET formation with a semitransparent Al mirror, and (b) photographs taken by a camera from the backside of the semitransparent Al mirror. (Collaboration with Takashi Kuma.)

The UV/blue-light write-beam exposure has a role to activate the hydrolytic condensation of the titanium oxide. The heating accelerates the hydrolytic condensation process, inducing an increasing in the refractive index in the exposed areas. Photographs of nanoscale optical circuits fabricated in a PRI sol-gel thin film are shown in Figure 6.12 [6]. The core thickness is ~230 nm and the core width is ~1 μm.

FIGURE 6.11 Dependence of coupling efficiency on writing time in the R-SOLNET with a semitransparent Al mirror. (Collaboration with Takashi Kuma.)

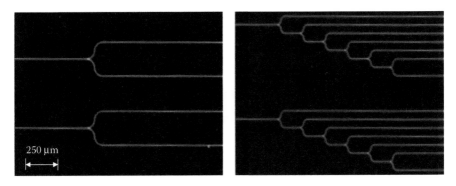

FIGURE 6.12 Top views of nanoscale optical circuits fabricated in PRI sol-gel thin films.

An experimental setup for formation of R-SOLNETs with reflective objects in the PRI sol-gel material is illustrated in Figure 6.13 [7]. On a Si substrate with a 2000-nm-thick SiO_2 film for an under-cladding layer, coating and prebaking of the PRI sol-gel material are carried out to deposit a PRI sol-gel thin film of approximately 230-nm thick. A silver paste droplet is placed as a reflective object in the PRI sol-gel thin film. A 405-nm write beam with power of 0.5 mW is introduced from a 3-μm optical fiber into the PRI sol-gel thin film under heating at 200°C to form an R-SOLNET. As Figure 6.14 shows, at writing time of 25 s, an R-SOLNET is slightly drawn to the silver paste droplet. At 50 s, the R-SOLNET is completely formed to be drawn to the droplet.

Figure 6.15 shows another example of R-SOLNET formation in the PRI sol-gel thin film [7]. A 405-nm write beam with power of 0.5 mW is introduced into the PRI sol-gel thin film for 30 s under heating at 200°C. Then, an R-SOLNET is formed toward a reflective defect, which acts as a reflective object in the PRI sol-gel thin film.

These results suggest that the R-SOLNET has a capability to find the locations of reflective objects and to guide light beams to the objects.

6.2.4 R-SOLNET with Luminescent Targets

As mentioned in Section 3.1, the R-SOLNET with luminescent targets has the following two advantages over the R-SOLNET with wavelength filters: the enhanced write-beam-overlapping effect (see Figure 3.2) and the controllability of the effective intensity balance of the write beam and luminescence (see Figure 3.3).

In this section, several demonstrations of the R-SOLNET with luminescent targets are presented.

6.2.4.1 Coumarin-481 Luminescent Targets

The first example is R-SOLNET with luminescent targets of coumarin-481 [8]. The luminescent targets are made of a laser dye, coumarin-481, which is dissolved in a photosensitive organic/inorganic hybrid material, SUNCONNECT® [9,10]. SUNCONNECT® is also used for the PRI material.

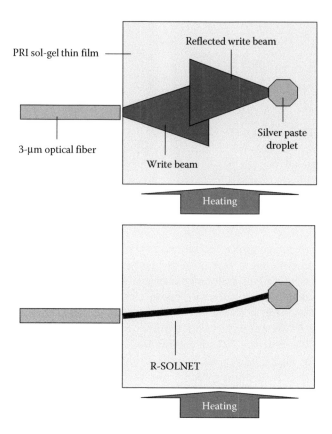

FIGURE 6.13 Experimental setup for formation of R-SOLNET with reflective objects in a PRI sol-gel thin film. (From S. Ono et al. *J. Lightwave Technol.* **27**, 5308–5313, 2009.) [7]

To enhance the targeting effect, the sensitivity of the PRI material to the luminescence should be comparable with, or higher than that of the write beam. As Figure 6.16 shows, in SUNCONNECT® the absorption spectrum has a peak in the UV region, and the absorption coefficient decreases at longer wavelengths, indicating that the sensitivity decreases with increasing the wavelength. Since the PL spectrum

FIGURE 6.14 Formation of an R-SOLNET between a 3-μm optical fiber and a silver paste droplet in a PRI sol-gel thin film. (From S. Ono et al. *J. Lightwave Technol.* **27**, 5308–5313, 2009.) [7]

Before SOLNET formation   After SOLNET formation

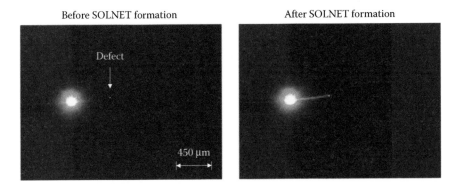

FIGURE 6.15 Formation of an R-SOLNET between a 3-μm optical fiber and a reflective defect in a PRI sol-gel thin film. (From S. Ono et al. *J. Lightwave Technol.* **27**, 5308–5313, 2009.) [7]

of coumarin-481/SUNCONNECT® has a peak around 470 nm, the wavelength separation between the PL peak and the write-beam wavelength is 65 nm for a 405-nm write beam, and 22 nm for a 448-nm write beam. The 448-nm write beam is therefore preferable for reducing the sensitivity unbalance between the write beam and the luminescence.

To deposit a luminescent target on the core region of an optical fiber edge, we used the light curing process shown in Figure 3.10a. The fiber edge was coated by a luminescent material of coumarin-481 dissolved in SUNCONNECT® at concentration of ~0.1 wt%. The coating was exposed to a UV light from the fiber core, curing the region just on the core. The luminescent target was obtained by wet-etching the coating to remove the noncured part using a mixture of isopropyl alcohol/4-methyl-2-pentanone. Figure 6.17 shows a photograph of a luminescent target emitting luminescence by excitation at 448 nm.

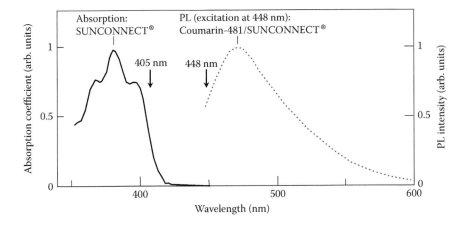

FIGURE 6.16 Absorption spectrum of SUNCONNECT® and PL spectrum of a coumarin-481/SUNCONNECT® luminescent target excited by a write beam at 448 nm.

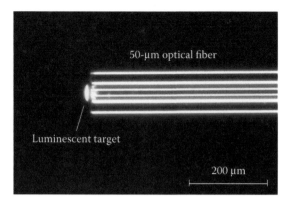

FIGURE 6.17 Photograph of a luminescent target under excitation at 448 nm. The target was deposited using the light curing process.

Figure 6.18 shows results of optical waveguide growth by a 405-nm write beam between a 50-μm fiber core and a cladding edge of an opposing optical fiber in SUNCONNECT®. The fiber-to-fiber gap distance is 250 μm and the lateral misalignment is 30 μm. The luminescent target of coumarin-481/SUNCONNECT® is on the core edge of the optical fiber on the right. The 405-nm write beam with power of 100 nW is introduced from the optical fiber on the left. A trace of optical waveguide growth, which is induced by the write beam from the left-hand-side optical fiber, develops. Another trace of optical waveguide growth, which is induced by the

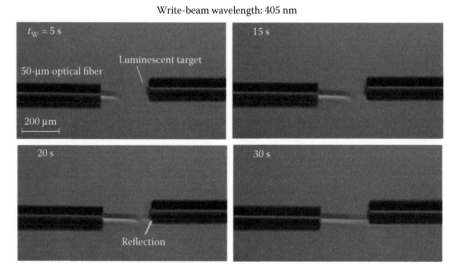

FIGURE 6.18 Optical waveguide growth between 50-μm optical fibers in SUNCONNECT® by a 405-nm write beam with power of 100 nW. An optical waveguide is grown between a 50-μm fiber core and the cladding edge of the opposing optical fiber.

reflected write beam from the edge of the right-hand-side optical fiber, also develops. The two traces finally merge.

As can be seen in Figure 6.18, emission from the luminescent target is not observed. Consequently, the luminescent target does not give rise to the "pulling water" effect necessary for formation of a self-aligned coupling waveguide toward the luminescent target. The absence of luminescence is explained as follows. Because the sensitivity of SUNCONNECT® is relatively high at 405 nm as can be seen in Figure 6.16, the power of the 405-nm write beam should be low (100 nW) for SOLNET formation according to the discussion in Section 5.3.2. This power is not sufficient to induce observable luminescence from the target.

Figure 6.19 shows an R-SOLNET formed by a 448-nm write beam between a 50-μm fiber core and a luminescent target on another 50-μm fiber core in SUNCONNECT®. The two optical fibers are placed with a gap distance of 250 μm and a lateral misalignment of 20 μm. A 448-nm write beam with power of 20 μW is introduced from the optical fiber on the left. Bright luminescence from the luminescent target is observed. As the write-beam irradiation progresses, the trace of optical waveguide growth from the core of the optical fiber on the left is observed. At the same time, the trace of optical waveguide growth from the luminescent target on the right is observed. The trace from the target is induced by luminescence from the target. The two traces finally merge to form a single self-aligned coupling waveguide to complete an R-SOLNET with luminescent targets between the optical fiber and the luminescent target.

Write-beam wavelength: 448 nm

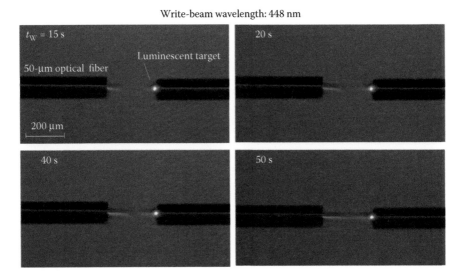

FIGURE 6.19 R-SOLNET formation between 50-μm optical fibers with 20-μm lateral misalignment in SUNCONNECT® by a 448-nm write beam with power of 20 μW. R-SOLNET is formed between a 50-μm optical fiber core and a luminescent target on another 50-μm optical fiber core.

Write-beam wavelength: 448 nm

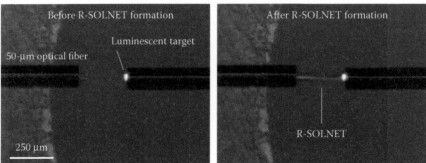

FIGURE 6.20 Another example of R-SOLNET formation between 50-μm optical fibers in SUNCONNECT® by a 448-nm write beam. (From T. Yoshimura *J. Lightwave Technol.* **33**, 849–856, 2015.) [11]

In Figure 6.20, another example of R-SOLNET formation between 50-μm optical fibers in SUNCONNECT® by a 448-nm write beam is presented [11]. A self-aligned coupling waveguide of R-SOLNET with luminescent targets is formed.

Figure 6.21 shows an R-SOLNET formed between a 9-μm fiber core and a luminescent target on another 9-μm fiber core [8]. A 448-nm write beam with power of 20 μW is introduced from the left-hand-side optical fiber. Luminescence emitted from the target is observed, and the trace of optical waveguide growth from the optical fiber on the left and another trace of optical waveguide growth from the

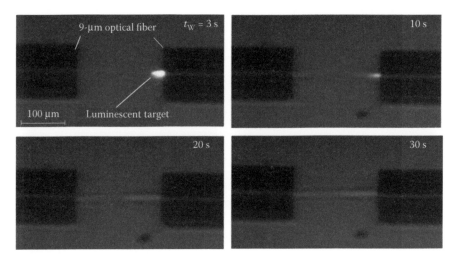

FIGURE 6.21 R-SOLNET formation between 9-μm optical fibers with 9-μm lateral misalignment in SUNCONNECT® by a 448-nm write beam with power of 20 μW. R-SOLNET is formed between a 9-μm optical fiber core and a luminescent target on another 9-μm optical fiber core.

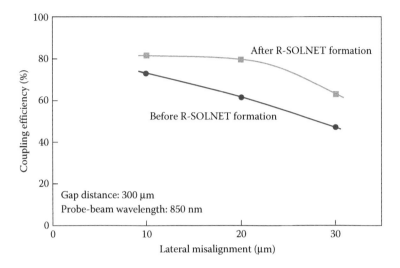

FIGURE 6.22   Dependence of maximum coupling efficiency on lateral misalignments in R-SOLNETs with luminescent targets between 50-μm optical fibers.

luminescent target are observed. The two traces finally merge to form a single self-aligned coupling waveguide to complete an R-SOLNET with luminescent targets.

The coupling efficiency between the 50-μm optical fibers connected by R-SOLNETs with luminescent targets was measured for a gap distance of 300 μm [8]. The probe-beam wavelength is 850 nm. As Figure 6.22 shows, the coupling efficiency decreases with the lateral misalignment between the optical fibers. By forming R-SOLNETs, the coupling efficiency increases by 10%–30%. The use of the R-SOLNET also increases tolerance to lateral misalignments. These results indicate that the R-SOLNET with luminescent targets has potential to realize the optical solder function for in-plane couplings, and, in principle, for surface-normal couplings such as optical Z-connections and couplings between waveguide gratings and optical fibers.

At present, the efficiency of the R-SOLNET coupling remains limited around 80%, which might be attributed to nonoptimized luminescent target structures. The target shape and the reproducibility of the target deposition should be improved by optimization of the fabrication processes.

6.2.4.2 Alq3 Luminescent Targets

Tris(8-hydroxyquinolinato) aluminum (Alq3) is known as phosphor for green emission in organic light emitting diodes. We attempted to use Alq3 for the luminescent target in R-SOLNETs.

The first experimental demonstration is shown in Figure 6.23 [12]. For the luminescent target, Alq3-dispersed polyvinyl alcohol (Alq3/PVA) fixed on an edge of an optical fiber is used. After a PRI material of NOA81/NOA65 (mixing weight ratio: NOA81:NOA65 = 1:2) was put dropwise on a slide glass, a 50-μm optical fiber and a luminescent target were placed there with a gap distance of ~2 mm. By a 405-nm

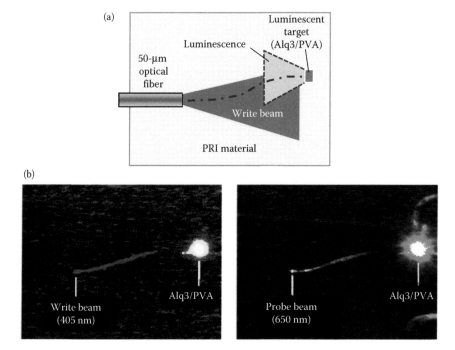

FIGURE 6.23 Formation of an R-SOLNET with luminescent targets toward an Alq3/PVA target. (From T. Yoshimura et al. *IEEE J. Select. Topics in Quantum Electron.* **18**, 1192–1199, 2012.) [12]

write beam introduced from the left-hand-side optical fiber into the PRI material, blue/green luminescence is emitted from the luminescent target. It is found that the 405-nm write beam is pulled to the luminescent target and a 650-nm probe beam is guided toward the target. This result indicates that the R-SOLNET with luminescent targets can be constructed in mm-scale sizes.

The second demonstration is shown in Figure 6.24 [13]. The experimental setup and materials are the same as those in the first demonstration. When a 405-nm write beam is introduced from the left-hand-side 50-μm optical fiber, initially, the write beam diffusely propagates in the PRI material. The Alq3/PVA target excited by the write beam emits blue/green luminescence. Then, the refractive index in the region, where the write beam and luminescence overlap, increases rapidly compared to the surrounding region. As a result, the write beam is guided toward the target as shown in Figure 6.24a, and an R-SOLNET that connects the fiber core and the luminescent target is formed as shown in Figure 6.24b. It is found that the R-SOLNET gradually expands from the fiber core diameter to the width of the Alq3/PVA target, constructing a tapered R-SOLNET. This indicates that the R-SOLNET with luminescent targets acts as optical solder with a function of a mode size converter.

In the third demonstration shown in Figure 6.25, an Alq3 thin film deposited on an optical fiber edge by the vacuum evaporation is used [14]. By introducing a 405-nm write beam from the left-hand-side 50-μm optical fiber into a PRI material

(a)

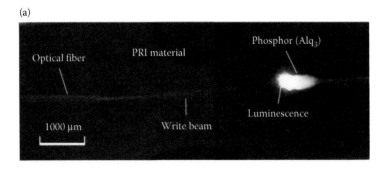

(b)

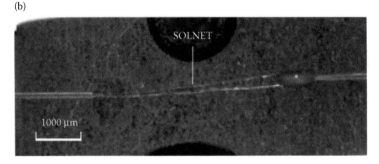

FIGURE 6.24 Formation of an R-SOLNET with luminescent targets between a 50-μm optical fiber and a large Alq3/PVA target. The R-SOLNET has a function of a mode size converter. (a) During write-beam exposure, (b) R-SOLNET with luminescent targets after 120-s write-beam exposure. (From M. Seki and T. Yoshimura, *Opt. Eng.* **51**, 074601–1–5, 2012.) [13]

of NOA81/NOA65, green/blue luminescence is generated from the Alq3 thin film (Figure 6.25b). The write beam is pulled toward the Alq3 thin film to form an R-SOLNET with luminescent targets (Figure 6.25c).

6.3 Y-BRANCHING SOLNETs

Figure 6.26 shows an experimental demonstration of R-SOLNET between a 50-μm optical fiber and two luminescent targets of coumarin-481/SUNCONNECT® on core edges of 50-μm optical fibers using a 448-nm write beam with power of 300 μW. An R-SOLNET with a Y-branching structure is observed in a PRI material of SUNCONNECT®.

Figure 6.27a shows another demonstration using Alq3/PVA luminescent targets and a 405-nm write beam with power of 6 μW in a PRI material of CV-sensitized NOA81/NOA65. With writing time, the write beam is initially pulled toward the lower target until 150 s. After that, the write beam begins to stretch toward the upper target, and finally, most of the write beam reaches the upper target. As shown in Figure 6.27b, a trace of a Y-branching R-SOLNET is observed. However, as can be seen in Figure 6.27c, most of the 650-nm probe beam propagates toward the upper target. This implies that although the R-SOLNET has a Y-branching structure, it does not work as a beam branching device.

(a)

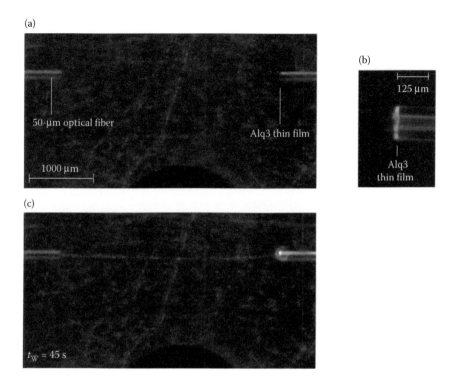

(b)

(c)

FIGURE 6.25 Formation of an R-SOLNET with luminescent targets between a 50-μm optical fiber and an Alq3 thin-film target deposited on an edge of an optical fiber. (a) Before write-beam exposure, (b) luminescence from Alq3 thin film, and (c) during write-beam exposure. (From T. Yoshimura and M. Seki, *J. Opt. Soc. Am. B* **30**, 1643–1650, 2013.) [14]

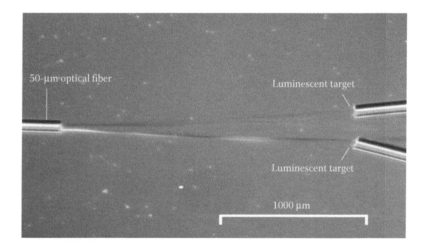

FIGURE 6.26 Formation of a Y-branching R-SOLNET from a 50-μm optical fiber toward two luminescent targets deposited on core edges of opposing 50-μm optical fibers. (Collaboration with Ryota Yajima.)

The sequential formation behavior of the Y-branching R-SOLNET shown in Figure 6.27a seems similar to that predicted by the FDTD simulation shown in Figure 4.83. So, it is speculated that the result shown in Figure 6.27a indicates the steeply-selecting characteristics of the SOLNETs, and the second branch toward the upper target is grown by a write beam leaked from the first branch toward the lower target.

(a)

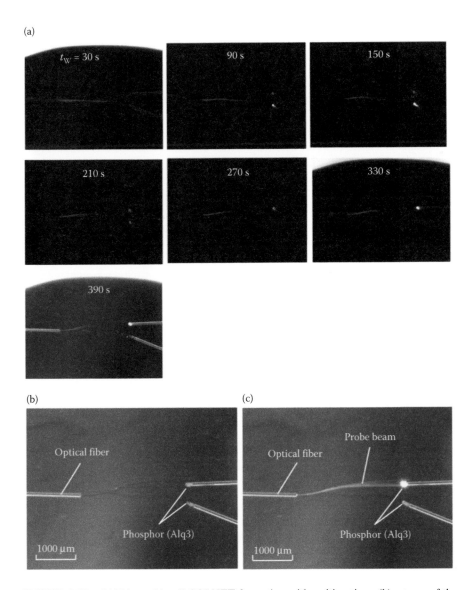

(b)   (c)

FIGURE 6.27 (a) Y-branching R-SOLNET formation with writing time, (b) a trace of the Y-branching R-SOLNET, and (c) 650-nm probe-beam propagation from the left to the right. (Collaboration with Keita Suzuki.)

6.4 TWO-PHOTON SOLNETs

6.4.1 Two-Photon TB-SOLNET

Experiments of two-photon TB-SOLNETs were performed in a two-photon PRI material of New-SUNCONNECT® containing two-photon sensitizers, CQ or BA. Here, New-SUNCONNECT® is a photosensitive organic/inorganic hybrid material, whose light absorption is smaller than the light absorption of SUNCONNECT® in blue wavelength regions. The small light absorption of the host material is preferable to feature the effect of the two-photon sensitizers.

Absorption spectra of CQ and BA are presented in Figure 6.28 for the $S_0 \rightarrow S_n$ and the $T_1 \rightarrow T_n$ transition, which are explained in Figure 3.8 [15]. Here, although only a $T_1 \rightarrow T_n$ absorption spectrum of BA is drawn, Brauchle et al. reported that the $T_1 \rightarrow T_n$ absorption spectrum of CQ is almost the same as that of BA. For the λ_1-write beam to induce the $S_0 \rightarrow S_n$ transition, a wavelength of 448 nm was chosen because CQ and BA absorb the 448-nm write beam as can be seen in Figure 6.28. For the λ_2-write beam, a wavelength of 850 or 780 nm, which is within the wavelength region for the $T_1 \rightarrow T_n$ transition, was chosen.

Figure 6.29a shows two-photon TB-SOLNET formation between 50-μm optical fibers in 20 wt%-CQ/New-SUNCONNECT® with a 448-nm write beam from the left and a 780-nm write beam from the right [16]. Green luminescence is emitted from CQ, which is excited by the 448-nm write beam. The intensity of the luminescence, which is probably accompanied with the $S_n \rightarrow S_0$ transition or the $T_1 \rightarrow S_0$ transition, becomes large with an increase in the 448-nm write-beam confinement. Therefore, the green luminescence can be used to monitor SOLNET formation dynamics. A line of green luminescence, which corresponds to an optical waveguide of two-photon TB-SOLNET, is observed between the optical fibers.

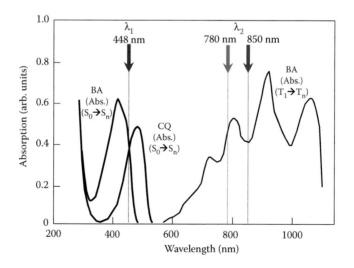

FIGURE 6.28 Absorption spectra of BA and CQ.

(a)

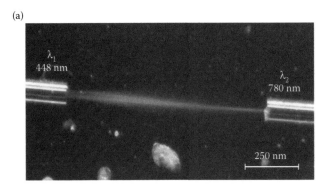

(b)

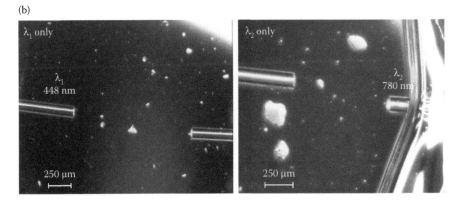

FIGURE 6.29 (a) Observation of two-photon TB-SOLNET formation monitored by PL from CQ and (b) photographs after only one of the two write beams are introduced in 20 wt%-CQ/New-SUNCONNECT®.

It should be noted that SOLNET formation is not observed when only one of the two write beams is introduced into the two-photon PRI material as shown in Figure 6.29b [17].

Figure 6.30 shows two-photon TB-SOLNET formation in BA/New-SUNCONNECT® using a 448-nm write beam and an 850-nm write beam [16]. The bending line of blue luminescence emitted from BA indicates that a self-aligned coupling waveguide of two-photon TB-SOLNET is formed between the two misaligned 50-μm optical fibers.

Two-photon TB-SOLNET formation between 50-μm optical fibers in 20 wt%-CQ/New-SUNCONNECT® with a 448-nm write beam from the left and an 850-nm write beam from the right is shown in Figure 6.31. A green luminescence line, which is initially deviated from the opposing right-hand-side fiber core, is gradually pulled toward the right-hand-side fiber core location with writing time. Finally, at 120 s the luminescence line connects the left-hand-side fiber core and the right-hand-side fiber core, indicating the two-photon TB-SOLNET formation. In the photograph, which was taken after the write beam turned off, a trace of the two-photon TB-SOLNET

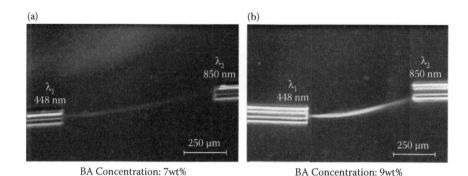

FIGURE 6.30 Observation of two-photon TB-SOLNET formation in BA/New-SUNCONNECT® monitored by PL from BA.

connecting the two optical fibers is observed. The TB-SOLNET keeps to connect the two optical fibers even after the left-hand-side optical fiber moves as Figure 6.32 shows.

In Figure 6.33, coupling efficiency of 448-nm probe beams that propagate from the left-hand-side optical fiber to the right-hand-side optical fiber is shown. By the two-photon TB-SOLNET formation, the coupling efficiency rises and the lateral misalignment tolerance is widened.

6.4.2 TWO-PHOTON R-SOLNET

For two-photon R-SOLNET formation, SUNCONNECT® containing a laser dye called styryl 7 or DCM is used for luminescent targets. In Figure 6.34a an absorption

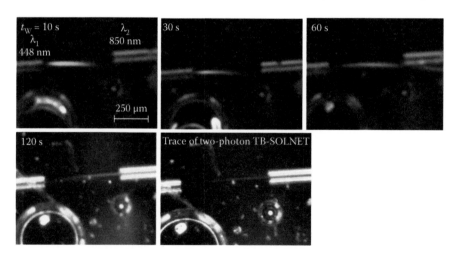

FIGURE 6.31 Two-photon TB-SOLNET formation between 50-μm optical fibers and a trace of the two-photon TB-SOLNET in 20 wt%-CQ/New-SUNCONNECT®. (Collaboration with Tomoya Hamazaki.)

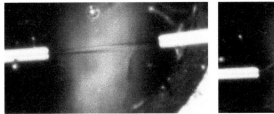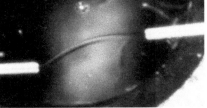

FIGURE 6.32 Photographs of the two-photon TB-SOLNET before and after the left-hand-side optical fiber is moved. (Collaboration with Tomoya Hamazaki.)

spectrum and a PL spectrum of styryl 7 in methanol [18] are shown with the $S_0 \to S_n$ absorption spectrum of CQ and the $T_1 \to T_n$ absorption spectrum of BA, which is almost the same as that of CQ [15]. Styryl 7 can absorb 448-nm write beam and emit luminescence in a wavelength range of 600–800 nm, which can be absorbed in the $T_1 \to T_n$ transition process. Therefore, styryl 7 is a suitable material for luminescent targets in the two-photon R-SOLNET formation. It is found from Figure 6.34b that DCM is another suitable material because DCM can absorb 448-nm write beam and emit luminescence, which can be absorbed in the $T_1 \to T_n$ transition process [18].

Formation of two-photon R-SOLNET with a luminescent target of 13 wt%-styryl 7/SUNCONNECT® between 50-μm optical fibers is shown in Figure 6.35. The two-photon PRI material is CQ/New-SUNCONNECT®. When a 448-nm write beam with power of 130 μW is emitted from the left-hand-side optical fiber, a green luminescence line monitoring the SOLNET formation is gradually pulled toward

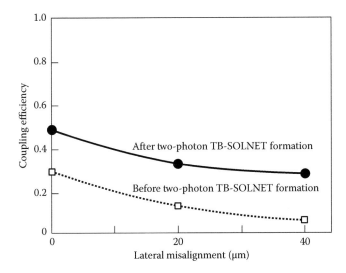

FIGURE 6.33 Dependence of maximum coupling efficiency on lateral misalignments in two-photon TB-SOLNETs formed between 50-μm optical fibers in 20 wt%-CQ/New-SUNCONNECT®. λ_1 and λ_2 are respectively 448 nm and 850 nm, and gap distance is 500 μm. (Collaboration with Tomoya Hamazaki.)

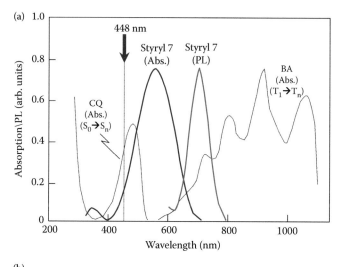

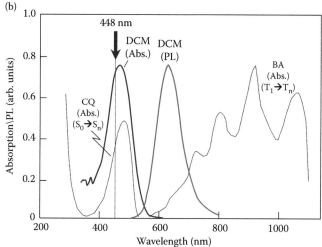

FIGURE 6.34 Absorption spectra of BA and CQ, and absorption/PL spectra of (a) styryl 7 and (b) DCM in methanol.

the luminescent target with writing time. Finally, the luminescence line reaches the luminescent target. A trace of the two-photon R-SOLNET connecting the two optical fibers is observed in Figure 6.36. These results confirm the two-photon R-SOLNET formation between the two optical fibers.

In Figure 6.37, coupling efficiency of 448-nm probe beams that propagate from the left-hand-side optical fiber to the right-hand-side optical fiber is shown. Although the efficiency is low at the present stage, a decrease in the coupling effect is suppressed by the SOLNET in a lateral misalignment range from 0 to ~20 μm. Optimization of the fabrication condition such as write-beam power and luminescent target structures might improve the efficiency.

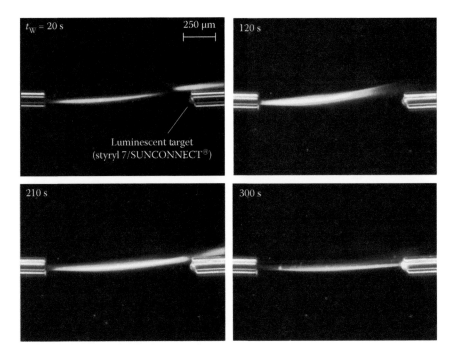

FIGURE 6.35 Two-photon R-SOLNET formation between 50-μm optical fibers in CQ/New-SUNCONNECT®. The luminescence target is styryl 7/SUNCONNECT®. 448-nm write beam is introduced from the left-hand-side optical fiber. (Collaboration with Shunya Yasuda, Hideaki Yamaura, and Yusuke Yamada.)

FIGURE 6.36 A trace of a two-photon R-SOLNET formed by a 448-nm write beam with power of 130 μW in 20 wt%-CQ/New-SUNCONNECT®. (Collaboration with Shunya Yasuda, Hideaki Yamaura, and Yusuke Yamada.)

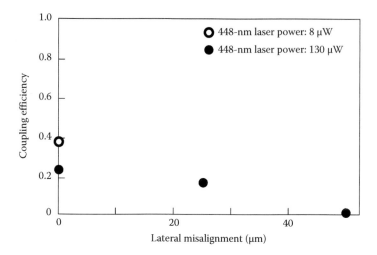

FIGURE 6.37 Dependence of maximum coupling efficiency on lateral misalignments in two-photon R-SOLNETs formed between 50-μm optical fibers in 20 wt%-CQ/New-SUNCONNECT®. The gap distance is 500 μm. (Collaboration with Shunya Yasuda, Hideaki Yamaura, and Yusuke Yamada.)

Figure 6.38 shows formation of a two-photon R-SOLNET with a luminescent target of 1 wt%-DCM/SUNCONNECT® between 50-μm optical fibers in CQ/New-SUNCONNECT®. The diameter of the right-hand-side optical fiber seems much larger than that of the left-hand-side optical fiber. This indicates that the right-hand-side optical fiber is located at a position closer to the objective lens. In other words, the two fibers are arranged with misalignments along both of the horizontal direction and the vertical direction. When a 448-nm write beam is introduced from the left-hand-side optical fiber, a green luminescence line is gradually pulled toward the luminescent target on the core edge of the right-hand-side optical fiber with writing

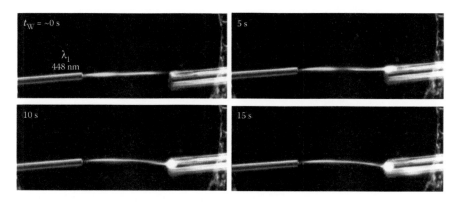

FIGURE 6.38 Two-photon R-SOLNET formation between 50-μm optical fibers with a large misalignment in CQ/New-SUNCONNECT®. The luminescence target is DCM/SUNCONNECT®. 448-nm write beam is introduced from the left-hand-side optical fiber. (Collaboration with Masataka Takashima and Riku Ito.) [19]

time, confirming that a two-photon R-SOLNET with luminescent targets is formed even when such a large misalignment exists.

REFERENCES

1. T. Yoshimura, J. Roman, Y. Takahashi, W. V. Wang, M. Inao, T. Ishitsuka, K. Tsukamoto, K. Motoyoshi, and W. Sotoyama, "Self-Organizing Waveguide Coupling Method "SOLNET" and its Application to Film Optical Circuit Substrates," *Proc. 50th Electronic Components & Technology Conference (ECTC)*, Las Vegas, Nevada, 962–969, 2000.
2. T. Yoshimura, J. Roman, Y. Takahashi, W. V. Wang, M. Inao, T. Ishituka, K. Tsukamoto, K. Motoyoshi, and W. Sotoyama, "Self-Organizing Lightwave Network (SOLNET) and Its Application to Film Optical Circuit Substrates," *IEEE Trans. Comp., Packag. Technol.* **24**, 500–509, 2001.
3. T. Yoshimura, A. Hori, Y. Yoshida, Y. Arai, H. Kurokawa, T. Namiki, and K. Asama, "Coupling Efficiencies in Reflective Self-Organized Lightwave Network (R-SOLNET) Simulated by the Beam Propagation Method," *IEEE Photon. Technol. Lett.* **17**, 1653–1655, 2005.
4. T. Yoshimura, T. Inoguchi, T. Yamamoto, S. Moriya, Y. Teramoto, Y. Arai, T. Namiki, and K. Asama, "Self-Organized Lightwave Network Based on Waveguide Films for Three-Dimensional Optical Wiring Within Boxes," *J. Lightwave Technol.* **22**, 2091–2100, 2004.
5. T. Yoshimura and H. Kaburagi, "Self-Organization of Optical Waveguides between Misaligned Devices Induced by Write-Beam Reflection," *Appl. Phys. Express* **1**, 062007, 2008.
6. S. Ono, T. Yoshimura, T. Sato, and J. Oshima, "Fabrication and Evaluation of Nano-Scale Optical Circuits using Sol-Gel Materials with Photo-Induced Refractive Index Variation Characteristics," *J. Lightwave Technol.* **27**, 1229–1235, 2009.
7. S. Ono, T. Yoshimura, T. Sato, and J. Oshima, "Fabrication of Self-Organized Optical Waveguides in Photo-Induced Refractive Index Variation Sol-Gel Materials with Large Index Contrast," *J. Lightwave Technol.* **27**, 5308–5313, 2009.
8. T. Yoshimura, M. Iida, and H. Nawata, "Self-Aligned Optical Couplings by Self-Organized Waveguides Toward Luminescent Targets in Organic/Inorganic Hybrid Materials," *Opt. Lett.* **39**, 3496–3499, 2014.
9. T. Sato, "Novel Organic-Inorganic Hybrid Materials for Optical Interconnects," *Proc. SPIE* **7944**, 79440M, 2011.
10. H. Nawata, "Organic-Inorganic Hybrid Material for on-Board Optical Interconnection and its Applications in Optical Coupling," *IEEE CPMT Symposium* Japan, 121–124, 2013.
11. T. Yoshimura, "Simulation of Self-Aligned Optical Coupling between Micro- and Nanoscale Devices Using Self-Organized Waveguides," *J. Lightwave Technol.* **33**, 849–856, 2015.
12. T. Yoshimura, C. Yoshino, K. Sasaki, T. Sato, and M. Seki, "Cancer Therapy Utilizing Molecular Layer Deposition and Self-Organized Lightwave Network: Proposal and Theoretical Prediction," *IEEE J. Select. Topics in Quantum Electron.* **18**, 1192–1199, 2012.
13. M. Seki and T. Yoshimura, "Reflective Self-Organizing Lightwave Network (R-SOLNET) Using A Phosphor," *Opt. Eng.* **51**, 074601-1-5, 2012.
14. T. Yoshimura and M. Seki, "Simulation of Self-Organized Parallel Waveguides Targeting Nanoscale Luminescent Objects," *J. Opt. Soc. Am. B* **30**, 1643–1650, 2013.
15. C. Brauchle, U. P. Wild, D. M. Burland, G. C. Bjorkund, and D. C. Alvares, "Two-Photon Holographic Recording with Continuous-Wave Lasers in the 750–1100-nm Range," *Opt. Lett.* **7**, 177–179, 1982.

16. T. Yoshimura and H. Nawata, "Micro/Nanoscale Self-Aligned Optical Couplings of the Self-Organized Lightwave Network (SOLNET) formed by Excitation Lights from Outside," *Opt. Commun.* **383**, 119–131, 2017.

17. T. Yoshimura, D. Takeda, T. Sato, Y. Kinugasa, and H. Nawata, "Polymer Waveguides Self-Organized by Two-Photon Photochemistry for Self-Aligned Optical Couplings with Wide Misalignment Tolerances," *Opt. Commun.* **362**, 81–86, 2016.

18. Catalogue of Laserdyes, Acros Organics.

19. T. Yoshimura, S. Yasuda, H. Yamaura, and Y. Yamada, "Self-Organized Lightwave Network (SOLNET) Formed by Two-Photon Photochemistry for 3-D Integrated Optical Interconnects," *2018 IEEE Optical Interconnects Conference (OI)*, Santa Fe, New Mexico, 2018 (submitted).

7 Applications of SOLNETs

Lightwaves have penetrated into communication/switching systems and computers to provide advanced OE systems, and into energy conversion and bio-medical fields. SOLNETs are expected to reduce the cost and energy dissipation and to create new architectures in these OE systems. SOLNETs are also expected to produce new types of solar cells, photosynthesis cells, photo-assisted cancer therapy including laser surgery and photodynamic therapy utilizing MLD, and so on. In this chapter, expected applications of SOLNETs are proposed.

7.1 INTEGRATED OPTICAL INTERCONNECTS AND SWITCHING SYSTEMS

Figure 7.1 illustrates a whole image of the lightwave penetration into communication/ switching systems and computers. Lightwaves were firstly introduced into long-haul communication systems, and then, they were implemented into shorter distance systems in the order of metro network, local-area network (LAN), and inter-box network for optical interconnects in data centers and super computers. And now, lightwaves are ready to be placed into boxes, reaching the final goal of the optical interconnects.

Within boxes of computers, increases in the clock frequency and the wiring density cause problems such as heat generation (power dissipation) and noise generation. Lightwave implementation will solve these problems. Lightwaves will also be required in massive switching systems to treat the rapidly-increasing data rates and volume in networking.

There are two remarkable features in the optical networking within boxes, being distinguished from the inter-box optical networking: (1) very short line distance of cm-mm order and (2) very large line count of hundreds or thousands.

The latter means that OE devices of hundreds or thousands should be distributed on a board or chip for electrical-to-optical (E-O) and optical-to-electrical (O-E) signal conversion. Currently, for the lightwave implementation, E-O/O-E conversion modules, in which discrete devices are assembled by packaging processes based on the flip-chip bonding, are used [1–4]. This approach causes the optics excess problem as follows.

- Excess components/materials
- Excess spaces for components
- Excess fabrication efforts
- Excess design efforts

The excess raises the system cost.

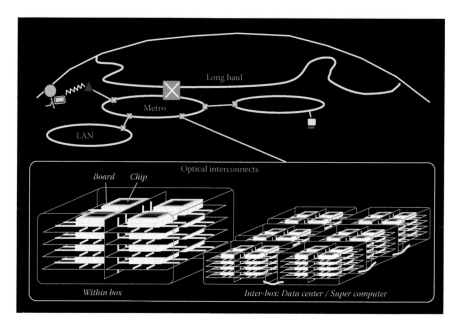

FIGURE 7.1 Whole image of the lightwave penetration into communication/switching systems and computers.

In this section, integrated optical interconnects and switching systems are proposed to solve the optics excess problem, and examples for "How SOLNETs contribute to constructing these proposed structures" are presented.

7.1.1 SCALABLE FILM OPTICAL LINK MODULE (S-FOLM)

To overcome the optics excess problem, the scalable film optical link module (S-FOLM) is expected to be a solution [5–11]. The concept is shown in Figure 7.2. Firstly, packaging elements such as electrical substrates and OE-Films are prepared. These are denoted as follows.

S: Electrical substrates.
W: Optical waveguide films.
D: Films, in which thin-film OE device flakes such as light modulators, optical switches (SWs), vertical cavity surface emitting lasers (VCSELs), photodetectors (PDs), and so on are embedded.
DW: Optical waveguide films, in which thin-film OE device flakes are embedded.
C: Films, in which thinned electronic chips such as large-scale integrated circuits (LSIs), integrated circuits (ICs), capacitors, and so on are embedded.

Then, the packaging elements are combined by the Film/Z-connection technology [12] to make products of the S-FOLM. Here, the Z-connections act as both electrical

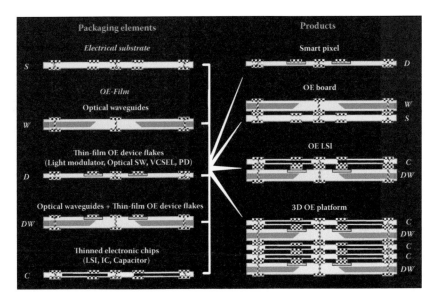

FIGURE 7.2 Concept of the scalable film optical link module (S-FOLM).

joints and mechanical joints between films. The S-FOLM realizes maximized variety of products made from minimized types of packaging elements, providing all levels of optical networking units. For example, an OE-Film (*D*) with thin-film VCSELs/PDs itself is a smart pixel. When an OE-Film (*W*) is attached on an electrical substrate (*S*), an OE board is constructed. A combination of an OE-Film (*DW*) and an OE-Film (*C*) with thinned LSIs provides an OE LSI. By stacking a plurality of OE-Films (*DW*) and OE-Films (*C*), a 3D OE platform is constructed. This implies that the S-FOLM has a great packaging scalability.

In the S-FOLM, small/thin divided OE device flakes are embedded just under the individual pad positions of LSI/IC chips. This gives the S-FOLM the following advantages to remove the optics excess and reduce the system cost.

- The material consumption for components and the fabrication efforts are minimized by using the device flake assembling process called "Photolithographic Packaging with Selectively Occupied Repeated Transfer (PL-Pack with SORT)" [8,11,13–15].
- Excess spaces for components are not necessary on the surface.
- Design efforts are minimized by inserting thinned interface ICs for flexible interface specifications.
- Bit error rates (BERs) and signal delays are reduced by replacing long electrical lines between the components with short vertical electrical lines.

In addition, the S-FOLM has a remarkable property of flexibility in structures and positions. The S-FOLM can be attached at any places with any shapes/sizes on substrates.

7.1.2   3D OE Platform Built by Self-Organized Optical Wiring

Figure 7.3 shows examples of the 3D OE platforms for integrated optical interconnects within boxes of computers and optical switching systems based on the S-FOLM [5–11]. The multilayer OE board shown in Figure 7.3a consists of a plurality of OE-Films (W) with vertical mirrors, an OE-Film (DW) with embedded thin-film light modulators (or VCSELs) and PDs, and an OE-Film (C). Vertical optical links between the optical waveguide films are carried out by optical Z-connections with vertical waveguides. The second layer, OE-Film (DW), is for the E-O and O-E conversion. The third and fourth layers, OE-Films (W), are for the multilayer optical signal routing. The optical signal link between 3D OE platforms are performed through the OE-Film (W) in the OE board by the backside connections.

The first layer, OE-Film (C), in which thinned interface ICs are embedded, is an interface film, on which LSIs are mounted. Via the interface film, electrical signals from output pads of LSIs are converted into optical signals by light modulators (or VCSELs). The optical signals are guided to PDs, where, the optical signals are converted to electrical signals for input pads of LSIs via the interface film. The interface film has functions such as driver/amplifier, MUX/DEMUX, error

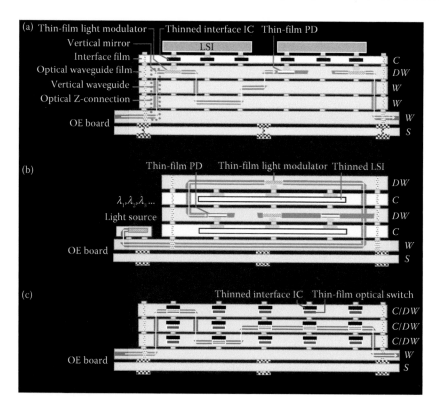

FIGURE 7.3   Concept of the 3D OE platforms for integrated optical interconnects within boxes of computers and optical switching systems based on the S-FOLM. (a) Multilayer OE board, (b) 3D-stacked OE MCM, and (c) 3D-MOSS.

correction, timing controlling, coding, and so on. This interface approach gives the 3D OE platforms a standardized-interface capability, and permits computer designers to build electrical architectures without knowledge of optics.

When the light modulators can directly be driven by LSI outputs and the PDs can directly generate electrical signals for LSI inputs as in the case of OE Amplifier/Driver-Less Substrates (OE-ADLES) [11,16], the interface film is not necessary.

In the 3D-stacked OE MCM shown in Figure 7.3b, light modulator/PD-embedded OE-Films (*DW*) are stacked together with OE-Films (*C*) containing thinned LSIs. Light sources are placed outside of the stacked structure. Optical signals, which are generated at E-O conversion sites in the OE-Films (*DW*) by LSI outputs, propagate to O-E conversion sites, being converted into electrical signals for LSI inputs. In some cases, the optical signals generated in the 3D-stacked OE MCM are coupled to optical waveguides in the OE-Film (*W*) of the OE board by the backside connections to be transmitted to other OE units, and vice versa.

In the 3D-stacked OE MCM, heat generation is a serious problem. So, implementation of OE-ADLES that uses external light sources of cw or pulse trains is effective for power dissipation reduction. By using external light sources with a plurality of wavelengths and by replacing the vertical mirrors with wavelength filters, wavelength division multiplexing (WDM) interconnects can be performed.

In the three-dimensional micro optical switching system (3D-MOSS) shown in Figure 7.3c, OE-Films (*DW*), in which thin-film optical switch arrays are embedded, are stacked with coupled OE-Films (*C*), where thinned interface ICs of optical switch drivers are embedded. Optical signals are successively switched by cascaded optical switches through 3D optical routing to be output. The 3D architecture is built by dividing conventional planar switching networks into a plurality of blocks and stacking them. The 3D structure is effective to reduce system sizes and wiring lengths, consequently, insertion loss. The details are described in Section 7.1.5.

In the 3D OE platforms, a large number of optical couplings in optical circuits are involved. In such cases, enormous alignment efforts with micron or submicron accuracy are required, raising system fabrication costs. Vertical waveguide formation for the 3D optical circuits is another issue.

In order to overcome these problems, a concept of the 3D self-organized optical wiring illustrated in Figure 7.4 is proposed. After a pre-form with connection units distributed three-dimensionally is prepared, optical waveguides of SOLNET are formed between the units automatically in a self-aligned manner by excitation, which generates write beams from the connection units, to construct the 3D self-organized optical wiring. By the SOLNET-based self-organized optical wiring, low-cost fabrication of the 3D OE platforms will be achieved.

7.1.3   SELF-ORGANIZED 3D-INTEGRATED OPTICAL INTERCONNECTS

The 3D-integrated optical interconnects with backbones of the self-organized optical wiring have a potential to achieve low-cost, small-size, and low-power dissipation optical interconnects [5–11]. The fabrication process is schematically depicted in Figure 7.5. Firstly, a pre-form shown in Figure 7.5a is prepared. OE-Films (*DW*) consisting of light modulators/PDs embedded by PL-Pack with SORT and nanoscale

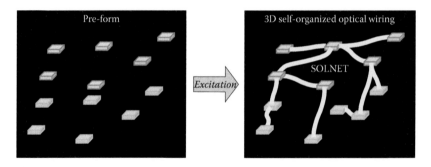

FIGURE 7.4   Concept of the 3D self-organized optical wiring of SOLNET.

waveguides with vertical mirrors are stacked together with OE-Films (*C*) containing thinned LSIs to construct a 3D stacked OE MCM on an OE board.

To form SOLNETs in the 3D structure, write-beam insertion is a serious concern. Because the 3D-integrated optical interconnects have a large number of optical coupling points, it is hard to introduce write beams into the points, particularly, points in the inner parts of the stacked structure. In such cases, R-SOLNETs, P-SOLNETs, and LA-SOLNETs are useful.

For in-plane couplings between nanoscale waveguides and light modulators with nanoscale waveguides, luminescent targets are deposited on their core edges and PRI materials are placed between them for LA-SOLNET formation. For vertical couplings between nanoscale waveguides in the 3D-stacked OE MCM and microscale waveguides in the OE board, phosphor-doped regions are distributed in these waveguides and PRI materials are placed between them for P-SOLNET formation. For inter-layer vertical couplings between nanoscale waveguides, phosphor-doped regions and luminescent targets are distributed for P/R-SOLNET formation. For couplings between optical fibers (or waveguides) and microscale waveguides in the OE board, luminescent targets are deposited at vertical mirror sites on the microscale waveguides for R-SOLNET formation. Similar optical coupling structures are available if the vertical mirrors are replaced with grating couplers. The luminescent target formation and the partial phosphor doping can be performed by the methods described in Section 3.4.

Then, excitation lights are introduced into the pre-form from outside. As illustrated in Figure 7.5b, self-organized in-plane couplings of LA-SOLNET between the nanoscale waveguides and the light modulators, self-organized vertical couplings of P-SOLNET between the nanoscale waveguides and the microscale waveguides, and self-organized inter-layer vertical couplings of P/R-SOLNET between the nanoscale waveguides in OE-Films are formed. For the couplings between the optical fibers (or waveguides) and the microscale waveguides, R-SOLNETs are formed by introducing write beams from the optical fibers (or waveguides). Thus, the self-organized 3D-integrated optical interconnects are built.

In the model shown in Figure 7.5, when the waveguide cores have emissive characteristics intrinsically, the LA-SOLNETs, P-SOLNETs, P/R-SOLNETs, and R-SOLNETs can be formed without the luminescent targets and the phosphor-doped regions. The SOLNETs are formed just by exposing the waveguide cores to excitation lights to generate write beams.

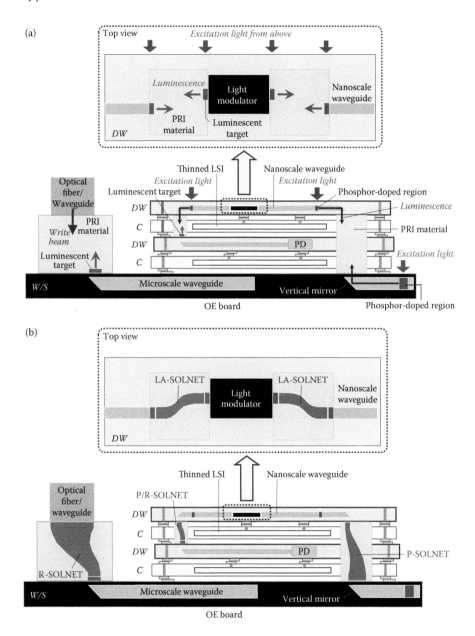

FIGURE 7.5 Process for self-organization of 3D-integrated optical interconnects. (a) Pre-form, (b) self-organization of 3D-integrated optical interconnects.

The technique utilizing SOLNETs is expected to realize low-cost optical couplings with wide spectral band widths because it enables the mode size conversion and the self-alignment simultaneously, and it is not wavelength-sensitive.

In addition, several examples of other coupling configurations with SOLNETs are briefly described [6,11]. For an optical waveguide-VCSEL (or PD) coupling, as shown

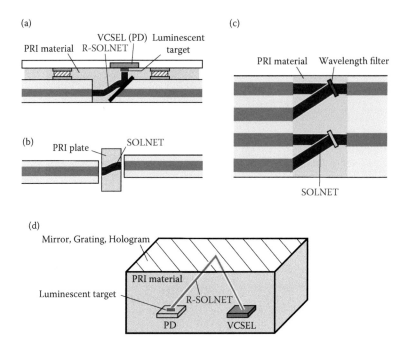

FIGURE 7.6 Additional coupling configurations with SOLNETs. (a) Optical waveguide-VCSEL (PD) coupling, (b) detachable optical solder using a PRI plate, (c) WDM coupling, and (d) Free-space coupling.

in Figure 7.6a, a luminescent target is deposited on the VCSEL (or PD). By introducing a write beam from the optical waveguide an R-SOLNET is formed between them. By inserting a PRI plate between optical devices as illustrated in Figure 7.6b, detachable optical solder of SOLNET can be made. By placing wavelength filters in a free space filled with a PRI material between optical waveguides and introducing write beams from the optical waveguides, as shown in Figure 7.6c, a plurality of SOLNETs are formed from the wavelength filters by partial reflection of the write beams at the wavelength filters. This enables us to insert wavelength filters into optical circuits without cutting or etching processes. Figure 7.6d shows an application of SOLNETs to free-space optical interconnects. In conventional systems, light beams propagate through relay lenses placed with sophisticated assembly. By implementation of R-SOLNETs, the relay lenses can be removed, realizing simple 3D optical wiring.

7.1.4 SELF-ORGANIZED 3D-MOSS

The 3D-MOSS has a potential to achieve high-speed massive optical switching with cost, size, and power dissipation reduction [8,11,17,18]. The fabrication process for self-organization of 3D-MOSSs is shown in Figure 7.7. Firstly, wavelength filters, which reflect write beams and pass signal beams, are deposited on core surfaces at the optical switch edges using ALD or MLD with processes mentioned in Section 3.4. Instead of the wavelength filters, luminescent targets can be used, too. Next, OE-Films are fabricated by embedding the optical switches into optical waveguide films using PL-Pack with

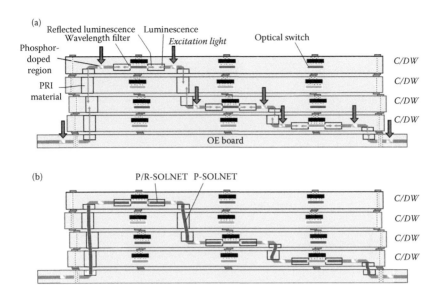

FIGURE 7.7   Process for self-organization of 3D-MOSS. (a) Pre-form, (b) self-organization of 3D-MOSS. (From T. Yoshimura et al. *IEEE J. Select. Topics in Quantum Electron.* **9**, 492–511, 2003.) [8]

SORT. In the optical waveguides, phosphor is doped for write-beam generation. If the optical waveguides themselves have emissive characteristics, the phosphor doping is not necessary. Then, as shown in Figure 7.7a, a pre-form is prepared by stacking the OE-Films with inserting PRI materials in parts, where SOLNETs will be formed.

By illuminating the phosphor-doped regions with excitation lights from outside, write beams are generated and propagate in the optical waveguides. As shown in Figure 7.7b, for connections between optical waveguides and optical switches, write beams introduced into the PRI materials from the optical waveguides are reflected by wavelength filters on the optical switch edges to form P/R-SOLNETs. For vertical waveguide construction, the write beams are introduced into the PRI materials from upper and lower optical waveguides through vertical mirrors to form P-SOLNETs. Thus, the self-organized 3D-MOSS is built.

7.1.5   Effect of SOLNET on Insertion Loss in 3D-MOSS

BPM calculations revealed that typical performance of a 1024 × 1024 3D-MOSS is as follows [8,11]; the system size is ~1.4 × 0.6 cm^2 and the insertion loss 29 dB for a 2^4-layer-stacked structure of OE-Films containing waveguide-prism-deflector micro-optical switches (WPD-MOSs). In this section, to demonstrate the advantage of SOLNET implementation into 3D OE platforms, the effect of SOLNETs on insertion loss in the 3D-MOSS is described [8,11,18].

Structural Model

Figure 7.8 shows a conventional 32 × 32 Banyan network consisting of five optical switch arrays [17]. Each array contains sixteen 2 × 2 optical switches. Adjacent

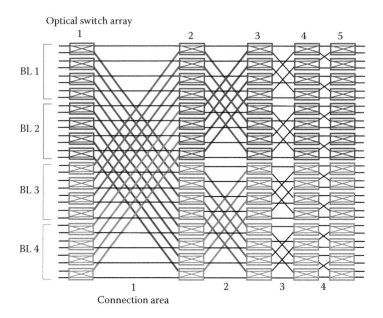

FIGURE 7.8 Schematic illustration of conventional 32 × 32 Banyan network. (From T. Yoshimura et al. *Opt. Eng.* **42**, 439–446, 2003.) [17]

arrays are connected by optical waveguides in connection areas. In connection area 1, an optical waveguide experiences 15 waveguide crossing points, 7 in connection area 2, 3 in connection area 3, and 1 in connection area 4.

To construct 3D-MOSS, the conventional Banyan network is divided into four sub-network blocks of BL 1, 2, 3, and 4 as shown in Figure 7.8. Inter- and intra-block connections are represented by thick and thin lines, respectively. As shown in Figure 7.9, the blocks are separated and stacked to each other to construct a 32 × 32 3D-MOSS. The long optical waveguides in connection area 1 are replaced with very-short-distance optical Z-connections, connecting BL 1–3 and BL 2–4 with experiencing no waveguide crossing points. The long optical waveguides in connection area 2 are also replaced with optical Z-connections. Thus, 3D-MOSS drastically reduces the waveguide crossing point count and the wiring distance.

Connection areas consisting of in-plane waveguides and corner-turning mirrors are denoted by "in-plane connection areas," and those containing optical Z-connections by "vertical connection areas." Connection areas 3 and 4 are in-plane connection areas, and connection areas 1 and 2 are vertical connection areas.

The advantage of 3D-MOSS becomes remarkable with increasing the channel count. For a 1024 × 1024 Banyan network, an optical waveguide in the connection area 1 experiences 511 crossing points in the conventional planar configurations. 3D-MOSS with eight sub-network blocks reduces the crossing point count to 0.

Table 7.1 summarizes structural parameters for a 1024 × 1024 3D-MOSS model, and a part of a 3D-MOSS is schematically shown in Figure 7.10. The channel count N_c of the 3D-MOSS is 1024. OE-Film thickness T_F is 4 μm and gap between the OE-Films

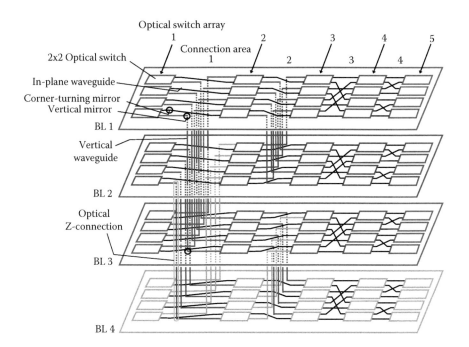

FIGURE 7.9 Schematic illustration of 32 × 32 3D-MOSS. (From T. Yoshimura et al. *IEEE J. Select. Topics in Quantum Electron.* **9**, 492–511, 2003.) [8]

TABLE 7.1
Structural Parameters for 1024 × 1024 3D-MOSS Model

Channel Count: N_C	1024
Layer Count: N_L	1, 2^2, 2^4, 2^6, 2^9
OE-Film Thickness/Gap: T_F/T_G	4 μm/46 μm
Optical Waveguide	
Width: w	4 μm
Refractive Index Clad: n_{Clad}	1.50
Core: n_{Core}	1.52
Pitch: P_{WG}	20 μm
2 × 2 WPD-MOS	
Size: Length L_{SW}, Width: W_{SW}	1190 μm, 100 μm
Step Count: S_{SW}	10
Pitch in an Array: P_{SW}	200 μm
Probe-Beam Wavelength	1.3 μm

Source: From T. Yoshimura et al. *IEEE J. Select. Topics in Quantum Electron.* **9**, 492–511, 2003. [8]

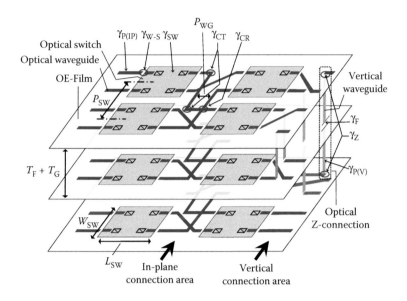

FIGURE 7.10 Schematic illustration of a part of 3D-MOSS. (From T. Yoshimura et al. *IEEE Photon. Technol. Lett.* **16**, 647–649, 2004.) [18]

T_G is 46 μm. So, thickness per one layer, $T_F + T_G$, is 50 μm. The 3D optical wiring consists of optical waveguides with a core width of 4 μm. The optical switches are 2×2 WPD-MOSs [8,11,19] with a length L_{SW} of 1190 μm and a width W_{SW} of 100 μm. The optical switch array count, that is, step count, S_{SW} is 10. Probe beams with a wavelength of 1.3 μm are used for calculations of coupling efficiency (or insertion loss).

When the network is divided into N_L blocks to make a 3D structure of N_L-layer-stacked OE-Films, the degree of 3D characteristics of the networks is expressed as follows.

$$D_{3D} = \log_2 N_L \tag{7.1}$$

Because D_{3D} indicates the number of vertical connection areas in a 3D-MOSS, connection areas 1 to D_{3D} are vertical connection areas, and connection areas $D_{3D} + 1$ to $S_{SW} - 1$ are in-plane connection areas. In the case of $N_L = 16$, namely, $D_{3D} = 4$, connection areas 1–4 and 5–9 are respectively vertical and in-plane connection areas.

As shown in Figure 7.10, in the present 3D-MOSS model, OE-Films are put with a pitch of $T_F + T_G$. In the OE-Films, 2×2 WPD-MOSs of length L_{SW} and width W_{SW} are arranged with a pitch of P_{SW}. Intra-layer wiring is carried out by in-plane crank-shaped waveguides with a pitch of P_{WG} while inter-layer wiring by optical Z-connections. Then, for in-plane connection area i, the connection area length $L(IP)_i$ and width $W(IP)$ are expressed as follows.

$$L(IP)_i = (N_C / 2^i + 1) P_{WG} \tag{7.2}$$

$$W(IP) = \left((N_C / 2) / 2^{D_{3D}} \right) P_{SW} \tag{7.3}$$

Here, in the expression for $L(IP)_i$, P_{WG} for buffer spaces is added.

For vertical connection area i, the connection area length $L(V)_i$, width $W(V)$, and height $H(V)$ are expressed as follows.

$$L(V)_i = (N_L/2^{i-1} + 5)P_{WG} \tag{7.4}$$

$$W(V) = ((N_C/2)/2^{D_{3D}})P_{SW} \tag{7.5}$$

$$H(V) = N_L(T_F + T_G) \tag{7.6}$$

Here again, in the expression for $L(V)_i$, P_{WG} for buffer spaces is included.

Then, total 3D-MOSS length $L_{3D\text{-}MOSS}$, width $W_{3D\text{-}MOSS}$, and height $H_{3D\text{-}MOSS}$, including switch array length, are given by,

$$L_{3D-MOSS} = L_{SW}S_{SW} + \left(\sum_{i=1}^{D_{3D}}(N_L/2^{i-1} + 5)\right)P_{WG} + \left(\sum_{i=D_{3D}+1}^{S_{SW}-1}(N_C/2^i + 1)\right)P_{WG}, \tag{7.7}$$

$$W_{3D-MOSS} = ((N_C/2)/2^{D_{3D}})P_{SW}, \tag{7.8}$$

$$H_{3D-MOSS} = N_L(T_F + T_G). \tag{7.9}$$

Insertion Loss

In Figure 7.10, major losses caused in 3D-MOSS are indicated by "γ." γ_P is propagation loss of optical waveguides, γ_{CR} loss induced at a right-angle waveguide–waveguide cross point, γ_Z loss of an optical Z-connection, γ_{CT} loss of a pair of corner turning mirrors in a crank-shaped waveguide, γ_F loss at a space in an OE-Film, γ_{SW} loss within a 2 × 2 optical switch, and $\gamma_{W\text{-}S}$ waveguide–switch coupling loss.

Major losses in 3D-MOSSs are listed in Table 7.2. γ_P is assumed 0.5 dB/cm, which is typical in fluorinated polyimide. γ_{CR} is estimated by the BPM simulation to be 0.0088 dB. γ_Z of 0.97 dB is estimated by the BPM/FDTD coupled simulation. Since a pair of corner turnings with two mirrors has the same structure as the optical Z-connection, γ_{CT} can also be 0.97 dB. A cross point of a vertical waveguide and a 4-μm-thick OE-Film can be regarded as a kind of a right-angle waveguide–waveguide cross point. So, γ_F can be 0.0088 dB. BPM calculation reveals that γ_F with a 25% misalignment is 0.89 dB. Loss for a TB-SOLNET/P-SOLNET coupling with a 25% misalignment is calculated to be 0.14 dB. Therefore, γ_F with 25% misalignment using the SOLNET is 0.14 dB, neglecting the SOLNET length dependence of γ_F. γ_{SW} is estimated to be 1.7 dB by the BPM simulation. $\gamma_{W\text{-}S}$ is basically 0 dB for no misalignment. For the case with a 25% misalignment, the BPM calculation reveals that $\gamma_{W\text{-}S}$ is 0.83 dB. For a SOLNET coupling with a 25% misalignment, similarly to the case of γ_F, $\gamma_{W\text{-}S}$ is 0.14 dB. In the present simulation, boundary reflection losses are neglected.

General expressions for the eight kinds of losses in 3D-MOSSs are derived in terms of the losses listed in Table 7.2. The results are summarized in Table 7.3. In-plane

TABLE 7.2

Major Losses in 3D-MOSSs

Propagation in Optical Waveguide: γ_P		0.5 dB/cm
Waveguide–Waveguide Cross Point: γ_{CR}		0.0088 dB
Pair of Corner Turnings (2 Corner-Turning Mirrors): γ_{CT}		0.97 dB
Optical Z-Connection: γ_Z		0.97 dB
Space in OE-Film: γ_F	with No Misalignment	0.0088 dB
	with 25% Misalignment	0.89 dB
	with 25% Misalignment using SOLNET	0.14 dB
2×2 Optical Switch: γ_{SW}		1.7 dB
Waveguide–Switch Coupling: γ_{W-S}	with No Misalignment	0 dB
	with 25% Misalignment	0.83 dB
	with 25% Misalignment using SOLNET	0.14 dB

Source: From T. Yoshimura et al. *IEEE J. Select. Topics in Quantum Electron.* **9,** 492–511, 2003. [8]

optical connections include in-plane propagation, waveguide–waveguide cross points, pairs of corner turnings, optical switches, and waveguide–switch couplings. Vertical optical connections include vertical propagation, optical Z-connections, and spaces in OE-Films. The expression for the loss of the space in OE-Film for SOLNET is obtained by assuming that each optical Z-connection is constructed by one SOLNET as depicted in Figure 7.7. The factor 2 in the expression is for two optical Z-connections between the 3D-MOSS and the OE board.

By using the parameters in Table 7.1, the length, width, and height of 3D-MOSSs can be respectively calculated from Equations 7.7, 7.8, and 7.9. Insertion losses of 3D-MOSSs can be obtained by summation of eight losses listed in Table 7.3.

In Figure 7.11, the occupation area, which is a product of the length and width of 3D-MOSSs, and the insertion loss of 3D-MOSSs are plotted as a function of the layer count N_L. $N_L = 1$ implies a conventional planar structure, and $N_L = 2^9$ implies that all the connection areas are vertical optical connections.

It is found that the occupation area decreases monotonically with N_L. For the insertion loss, on the other hand, it has a peak in the region of $N_L = 2^4 \sim 2^6$ when no misalignments exist in the 3D-MOSS. Insertion loss at $N_L = 1$ is 43 dB. With an increase in N_L, the insertion loss decreases to 29 dB at $N_L = 2^4$, then, increases reaching 40 dB at $N_L = 2^9$. This result indicates that a certain optimum layer count exists for minimizing insertion loss of 3D-MOSSs. The dashed curve represents insertion loss for 25% misalignment case, where the misalignment exists between adjacent layers and between an in-plane waveguide and a waveguide in a WPD-MOS (note that the dashed curve is drawn with 1/24 reduction). The misalignment causes a drastic monotonic increase in insertion loss with increasing N_L. By implementing SOLNETs

TABLE 7.3
General Expressions for Losses in 3D-MOSSs

In-Plane Optical Connections

In-Plane Propagation

$$\left[\left(\sum_{i=1}^{D_{3D}}(N_L/2^{i-1}+5)\right)P_{WG} + \left(\sum_{i=D_{3D}+1}^{S_{SW}-1}(N_C/2^i+1)\right)P_{WG} + \left(\sum_{i=D_{3D}+1}^{S_{SW}-1}(N_C/2)/2^i\right)P_{SW}\right]\gamma_P$$

Waveguide–Waveguide Cross Point

$$\left[\sum_{i=D_{3D}+1}^{S_{SW}-1}N_C/2^i\right]\gamma_{CR}$$

Pair of Corner Turnings

$$[S_{SW}-1-D_{3D}]\gamma_{CT}$$

Optical Switch

$$[S_{SW}]\gamma_{SW}$$

Waveguide–Switch Coupling

$$[2S_{SW}]\gamma_{W-S}$$

Vertical Optical Connections

Vertical Propagation

$$[2N_L(T_F+T_G)]\gamma_P$$

Optical Z-Connection

$$[2+D_{3D}]\gamma_Z$$

Space in OE-Film

$$[2N_L]\gamma_F \quad [2+D_{3D}]\gamma_F \text{ for SOLNET}$$

Source: From T. Yoshimura et al. *IEEE J. Select. Topics in Quantum Electron.* **9**, 492–511, 2003. [8]

into 3D-MOSSs with 25% misalignment, the insertion loss decreases, reaching a level comparable to that of 3D-MOSSs with no misalignments.

For small N_L, a main contribution to the loss is waveguide–waveguide cross points while spaces in OE-Films for large N_L [18]. The results indicate that the insertion loss of the 3D-MOSS can be minimized by selecting N_L.

It was reported that high-accuracy bonding equipment enables layer stacking with a positional accuracy of 1 μm for 1 cm² chips [20]. Therefore, for 3D-MOSSs of 1.424 cm × 0.64 cm, it is consistent to assume an acceptable positional misalignment of ~1 μm that corresponds to 25% misalignment for optical circuit backbones consisting of 4-μm-wide optical waveguides.

The insertion loss of 32 dB for the 1024 × 1024 SOLNET-implemented 3D-MOSS is still high. For further simplification of the system, each loss listed in Table 7.2 should be decreased by optimizing individual structures and materials.

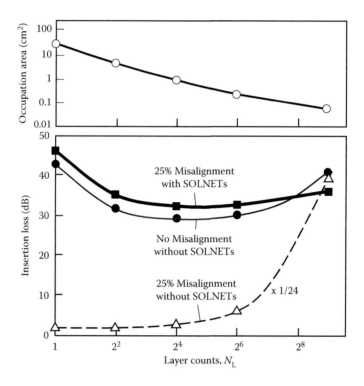

FIGURE 7.11   Occupation area and insertion loss of 3D-MOSSs plotted as a function of the layer count. (From T. Yoshimura et al. *IEEE Photon. Technol. Lett.* **16**, 647–649, 2004.) [18]

7.2   INTEGRATED SOLAR ENERGY CONVERSION SYSTEMS

7.2.1   Waveguide-Type Thin-Film Sensitized Solar Cell

The sensitized solar cell has received much attention as a cost-effective next-generation solar energy conversion module [21]. In conventional cells, porous semiconductors are used to increase the number of adsorbed dye molecules on the semiconductor surfaces to enhance light absorption. In such structures, however, the internal resistivity of the cells becomes large because the porous structures decrease the transport channel width and the mobility for electrons.

To avoid this problem, it is preferable to use thin-film semiconductors with high-crystalline quality in the cells. In such structures, however, normally-incident light beams pass through very thin sensitizing layers, resulting in very small light absorption. The waveguide-type thin-film sensitized solar cell [22–26], in which light beams propagate in thin-film semiconductors with the guided light configuration to enhance light absorption, is expected to solve the problem.

In the waveguide-type cell it is important to collect light beams and introduce them into the thin-film semiconductors efficiently. To do this, we proposed the integrated solar energy conversion system with embedded waveguide-type thin-film sensitized solar cells schematically shown in Figure 7.12 [26–28].

(a)

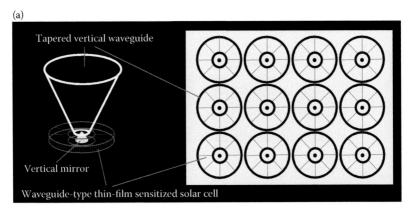

(b)

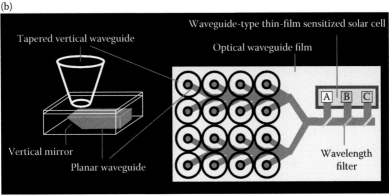

FIGURE 7.12 Concept of the integrated solar energy conversion system with embedded waveguide-type thin-film sensitized solar cells. Incident light beams are collected by (a) tapered vertical waveguides and (b) tapered vertical waveguides and planar waveguides.

In the system shown in Figure 7.12a, incident light beams collected by tapered vertical waveguides are guided to the bottom parts of the waveguides, where cone-shaped vertical mirrors are placed, to be introduced into the embedded solar cells via the vertical mirrors.

In the system shown in Figure 7.12b, a film with tapered vertical waveguides and an optical waveguide film with planar waveguides having vertical mirrors are stacked. Incident light beams collected by the tapered vertical waveguides are introduced into the planar waveguides via the vertical mirrors, and are guided to the embedded solar cells. By embedding wavelength filters into the optical waveguide film, light beams can branch to semiconductors having matched spectral responses, for example, UV-blue beams to ZnO, visible beams to ZnO sensitized in the visible wavelength region, and infrared beams to ZnO sensitized in the infrared wavelength region.

These configurations of the integrated solar energy conversion system enable reduction of the internal resistivity of the solar cells. They also enable reduction of the semiconductor consumption [11,26–28].

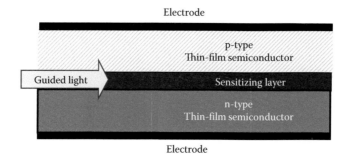

FIGURE 7.13 Waveguide-type thin-film sensitized solar cell with a structure of n-type semiconductor/sensitizing layer/p-type semiconductor.

Figure 7.13 shows an example of the waveguide-type thin-film sensitized solar cell with a structure of n-type thin-film semiconductor/sensitizing layer/p-type thin-film semiconductor [22,26]. High-crystalline-quality thin-film semiconductors with high electron mobility decreases the internal resistivity to improve the photovoltaic performance. Light beams are guided in the cell. In such guided light configuration, light beams pass through a lot of sensitizing agents in the sensitizing layer to enhance photocurrents. Thus, high-performance sensitized solar cells are expected.

In preliminary experiments, photocurrent enhancement by a factor of 4~10 was observed in the guided light configuration compared to the conventional normally-incident light configuration [22]. In addition, the guided light configuration enables us to replace expensive and electrically-resistive transparent electrodes with low-cost and low-resistivity metal electrodes.

The sensitizing layer can be organic multiple quantum dots (MQDs), namely, molecular MQDs and polymer MQDs that are grown on the thin-film semiconductor surfaces by MLD [22–26]. Figure 7.14 shows an example of molecular-MQD sensitization with a light-harvesting antenna consisting of linearly-connected

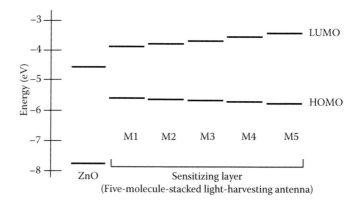

FIGURE 7.14 Molecular-MQD sensitization with a light-harvesting antenna consisting of linearly-connected five molecules.

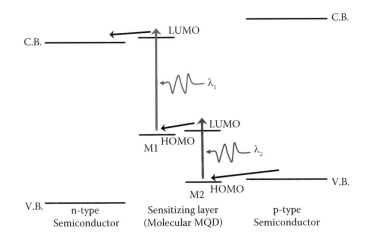

FIGURE 7.15 Improved molecular-MQD sensitization utilizing the two-step electron excitation.

five molecules; M1–M5. Excited electrons in the molecules are injected into the semiconductor (ZnO) through the antenna with a gradual energy slope [25,26].

Figure 7.15 shows improved molecular-MQD sensitization utilizing the two-step electron excitation [24,26] that is similar to the z-scheme process in the photosynthesis of plants. An electron excited by a photon with a wavelength of λ_1 in molecule M1 is injected into the n-type semiconductor. An electron excited by a photon with a wavelength of λ_2 in molecule M2 is transferred to HOMO of M1. The hole left in HOMO of M2 is compensated by an electron from the p-type semiconductor. Thus, electrons travel cyclically as—[p-type semiconductor] → [HOMO/M2] → [LUMO/M2] → [HOMO/M1] → [LUMO/M1] → [n-type semiconductor] → [p-type semiconductor] →. This sensitization mechanism suppresses the energy loss arising from the excess photon energy in light absorption processes [23,24,26], and at the same time, it increases the energy difference between the Fermi level of the n-type semiconductor and the HOMO of the sensitizing M2 molecules to increase the voltage generated from the solar cell [24,26].

7.2.2 WAVEGUIDE-TYPE THIN-FILM ARTIFICIAL PHOTOSYNTHESIS CELL

To date, artificial photosynthesis of H_2 and O_2 has been achieved [29], and photosynthesis of formate utilizing semiconductor/complex hybrid photocatalysts has also been reported [30]. The integrated solar energy conversion system is expected to be applied to the artificial photosynthesis [26].

Figure 7.16a shows an example of the thin-film photosynthesis cell, which is in the integrated solar energy conversion system using 3D optical circuits with embedded wavelength filters [26]. Light beams branch into two lightwave paths for wavelengths of λ_1 and λ_2, and the two light beams are respectively introduced onto two thin-film semiconductors in water, generating H_2 and O_2 utilizing the z-scheme process in the photosynthesis of plants. The photosynthesis process using two kinds of

(a)

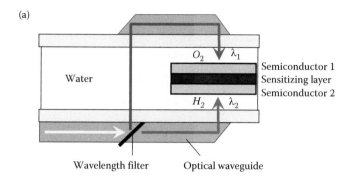

(b)

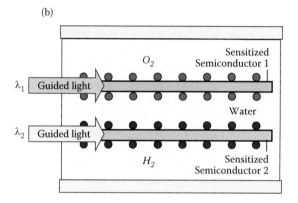

FIGURE 7.16 Concept of the photosynthesis cells with thin-film semiconductors. (a) Thin-film photosynthesis cell, (b) Waveguide-type thin-film sensitized photosynthesis cell.

semiconductors was developed by Sayama and Arakawa [29] using water, in which two kinds of semiconductor powders are dispersed. In the powder-based process, however, the semiconductor positions are difficult to be controlled. The thin-film photosynthesis seems promising because the relative positions between the two kinds of thin-film semiconductors can be optimized.

In the waveguide-type thin-film sensitized photosynthesis cell shown in Figure 7.16b, guided light beams in the thin-film semiconductors pass through a lot of sensitizing agents. The guided light configuration enhances the light absorption of the thin-film semiconductors to generate gases efficiently [26].

In the photosynthesis of H_2, high-power-density light-beam exposure is required to achieve high conversion efficiency because the H_2O oxidization is induced by the four-electron process. The waveguide-based system is preferable to increase the light-beam power density. Therefore, the integrated solar energy conversion system with the guided light configuration is expected to improve the conversion efficiency in the photosynthesis.

When the improved molecular-MQD sensitization with two-step electron excitation shown in Figure 7.15 is applied to the water splitting, H_2 would be generated from the n-type semiconductor/water interface (left-hand side in Figure 7.15), where

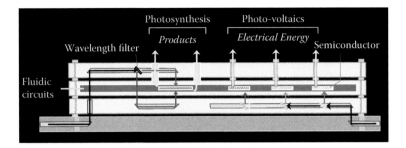

FIGURE 7.17   Concept of the integrated photonic/electronic/chemical system (IPECS).

cocatalyst is deposited in some cases, and O_2 from the p-type semiconductor/water interface (right-hand side in Figure 7.15). Meanwhile, it is known that light-harvesting antennas raise density of charge carriers for the H_2O oxidization. The sensitization utilizing light-harvesting antennas grown by MLD might enhance the photosynthesis efficiency further [25].

A concept of the integrated photonic/electronic/chemical system (IPECS) is shown in Figure 7.17 [26]. IPECS contains fluidic circuits for supplying liquids such as electrolyte, water, and various solvents to the system. Embedded wavelength filters in IPECS enable efficient artificial photosynthesis as well as photovoltaics. Since the structure of the IPECS is similar to that of the 3D-Stacked OE MCM shown in Figure 7.3b, the fabrication process is similar.

7.2.3   SELF-ORGANIZED INTEGRATED SOLAR ENERGY CONVERSION SYSTEM

A concern in the integrated solar energy conversion systems is the collection of free-space light beams into planar waveguides in the optical waveguide films, and the coupling of the light beams from the planar waveguides into the thin-film semiconductors in waveguide-type thin-film solar cells or photosynthesis cells. The optical solder of SOLNET might provide the solution as illustrated in Figure 7.18.

For the free-space light-beam collection into the planar waveguides, tapered vertical waveguides self-aligned to the vertical mirrors are constructed using R-SOLNETs as follows. An array of luminescent targets or wavelength filters are deposited at the vertical mirror sites on the optical waveguide film. After a PRI material layer is coated on the film surface, R-SOLNETs are formed by blanket write-beam exposure to fabricate an array of the tapered vertical waveguides.

For optical couplings between the planar waveguides and the thin-film semiconductors, luminescent targets or wavelength filters are deposited on the edges of the thin-film semiconductors. After a PRI material is inserted between the planar waveguides and the thin-film semiconductors, self-aligned coupling waveguides of R-SOLNET are formed by introducing write beams from the planar waveguides to the PRI material to fabricate optical solder.

In the case of IPECSs, similar SOLNET-based coupling methods for the 3D self-organized optical wiring shown in Figures 7.5 and 7.7 can be applied.

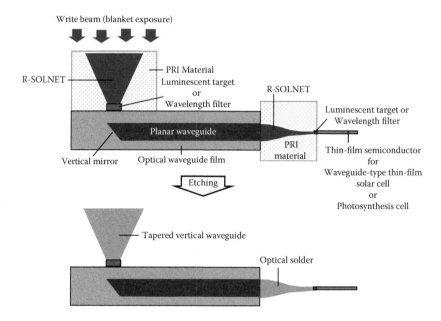

FIGURE 7.18 SOLNET-implemented integrated solar energy conversion system.

7.3 PHOTO-ASSISTED CANCER THERAPY

7.3.1 SOLNET-Assisted Laser Surgery

The R-SOLNET with luminescent targets enables optical waveguide formation toward microscale or nanoscale luminescent targets. This suggests a possibility of SOLNET-assisted laser surgery [22,26,31,32].

Figure 7.19 shows a schematic illustration for the concept of the SOLNET-assisted laser surgery. Firstly, luminescent molecules are connected to cancer cells by the liquid-phase MLD (LP-MLD) described in Appendix II, where the human body is regarded as an MLD chamber and cancer cells as substrates. In the first step of the LP-MLD process, anchoring molecules that have a capability to be connected to cancer cells selectively are injected into the body to be adsorbed by the cancer cells. In the second step, after the excessive anchoring molecules are excreted, luminescent molecules that have a capability to be connected to the anchoring molecules selectively are injected to make the cancer cells luminescent targets. If necessary, further sequential molecule injections are carried out.

Next, an optical fiber and a PRI material are placed in a region surrounding the cancer cells, and then, a write beam is introduced from the optical fiber. The luminescent molecules absorb the write beam and generate luminescence to form R-SOLNETs that connect the optical fiber to the cancer cells. Finally, by introducing surgery laser beams into the R-SOLNETs via the optical fiber, cancer cells are destroyed selectively. By detecting the backward luminescence emitted from the luminescent molecules, *in situ* monitoring of the degree of the cancer destruction might be possible.

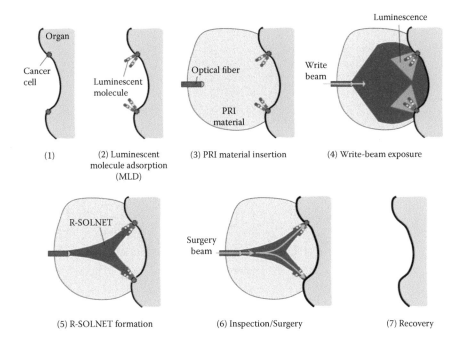

FIGURE 7.19   Concept of the SOLNET-assisted laser surgery.

Figure 7.20 shows a schematic illustration for the concept of the SOLNET-assisted laser surgery using an optical waveguide array [32]. A plurality of write beams are introduced in parallel from optical waveguides in the array to form R-SOLNETs toward the nearest cancer cells with adsorbed luminescent molecules. In this case, branching of SOLNETs is not necessary, enabling the straightforward surgery beam guiding to the cancer cells.

For practical applications of SOLNETs to the laser surgery, it is necessary to develop appropriate nontoxic PRI materials and anchoring molecules with functions to be selectively adsorbed by cancer cells. By using two-photon R-SOLNETs, the targeting ability is expected to be improved.

Simulations of the SOLNET-assisted laser surgery were performed for several models using the FDTD method described in Chapter 4. Figure 7.21 shows simulation results for a model involving a 600-nm wide luminescent target of a cancer cell site with adsorbed luminescent molecules [31]. The target is located with a lateral misalignment of 600 nm from the axis of a 1.2-μm-wide and 3-μm long waveguide core. Efficiency for the luminescence generation from the target is assumed 0.7. The refractive index of the waveguide core is 1.8 and that of the cladding region is 1.0. The refractive index of the PRI material increases from 1.5 to 1.7 upon write-beam exposure.

In order to avoid the light absorption by hemoglobin in human bodies, 650 nm is chosen for the wavelength of the write beam and the surgery beam because hemoglobin has a high-transmittance window in a wavelength region longer than $\sim$600 nm. Since wavelengths of luminescence are longer than the excitation light wavelengths due to the Stokes shift, 700 nm is chosen for the luminescence wavelength.

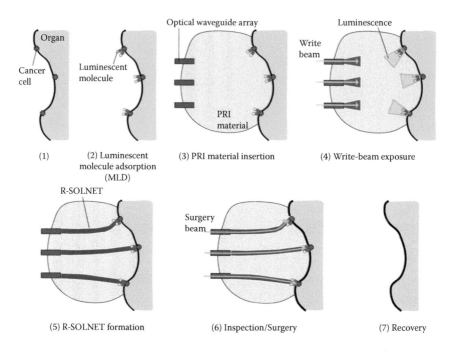

FIGURE 7.20 Concept of the SOLNET-assisted laser surgery using an optical waveguide array.

In Figure 7.21, the upper and lower figures respectively represent n^2 and intensity of surgery beams E^2 (surgery beam). When a write beam is emitted from the waveguide core on the left, with writing time, a trace of optical waveguide growth by the write beam appears from the left-hand side, and at the same time, another trace of optical waveguide growth by the luminescence emitted from the target appears from the right-hand side. Finally, the write beam and luminescence form a one-photon R-SOLNET between the waveguide core and the target by self-focusing. Accordingly, a surgery beam is guided toward the target of the cancer cell site.

Coupling efficiency of the surgery beam to the cancer cell site is plotted as a function of writing time in Figure 7.22. Efficiency over 90% is expected after 4.8-s writing. These results reveal the prospect of the SOLNET-assisted laser surgery for cancer.

Figure 7.23 shows results for FDTD simulations of the SOLNET-assisted laser surgery in the case that two luminescent targets of cancer cell sites are involved [31]. It is found that luminescence is generated from the two targets by the write-beam exposure to form a branching one-photon R-SOLNET toward the two targets. Accordingly, surgery beams are guided toward the targets. This result suggests that the R-SOLNET has a capability to catch multiple targets of cancer cell sites.

Figure 7.24 shows results for FDTD simulations of the SOLNET-assisted laser surgery using an optical waveguide array [32]. When write beams are emitted from the waveguide cores on the left, luminescence is emitted from luminescent targets of cancer cell sites to form one-photon R-SOLNETs between the waveguide cores and

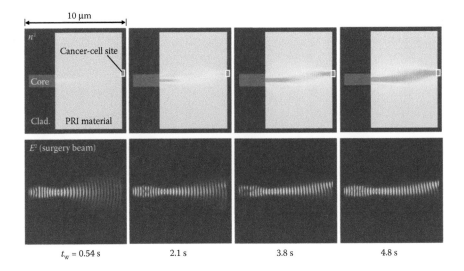

FIGURE 7.21   FDTD simulations of the SOLNET-assisted laser surgery.

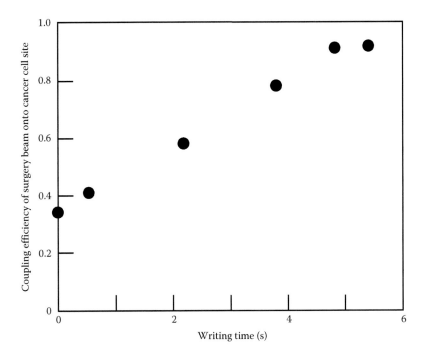

FIGURE 7.22   Coupling efficiency of the surgery beam to the cancer-cell site as a function of writing time.

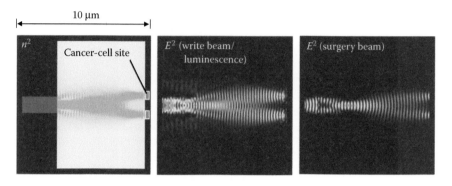

FIGURE 7.23 FDTD simulations of the SOLNET-assisted laser surgery in the case that two cancer-cell sites are involved.

the nearest targets. When surgery beams are introduced from the waveguide cores, they are guided to the nearest targets of the cancer cell sites.

In the present simulations, the area of the models is $10 \times 10 \ \mu m^2$ or $32 \times 32 \ \mu m^2$ and the cancer cell size is 600 nm. The model sizes can be rescaled. For example, by 1.7×10^3–rescaling, the area of $10 \times 10 \ \mu m^2$ and the cancer size of 600 nm are enlarged to $1.7 \times 1.7 \ cm^2$ and 1 mm, respectively.

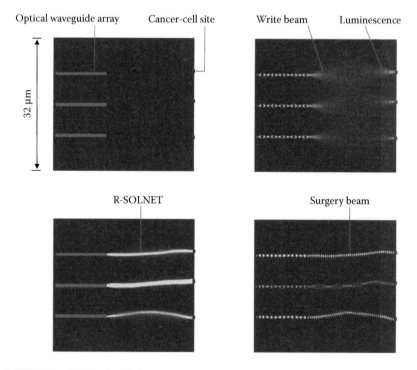

FIGURE 7.24 FDTD simulations of the SOLNET-assisted laser surgery with an optical waveguide array.

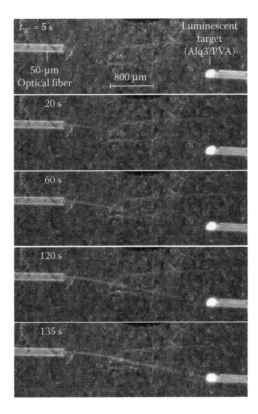

FIGURE 7.25   R-SOLNET formation between a 50-μm optical fiber and a luminescent target of Alq3-dispersed PVA in a PRI material containing CV. (Collaboration with Masaya Sato.)

An experimental result of self-aligned coupling waveguide formation toward a luminescent target is shown in Figure 7.25 [32]. A PRI material of NOA81/NOA65 sensitized by 0.1-wt% CV is used. A target of Alq3/PVA is deposited on an optical fiber edge. When a 405-nm write beam is emitted from a 50-μm optical fiber, green/blue luminescence is generated from the luminescent target. At the same time, red luminescence from CV doped in the PRI material is observed. The red luminescence enables us to trace R-SOLNET formation. With writing time, a one-photon R-SOLNET is formed toward the target, and finally the optical fiber and the target are connected.

7.3.2   SOLNET-ASSISTED PHOTODYNAMIC THERAPY

Figure 7.26 shows a schematic illustration for the concept of the SOLNET-assisted photodynamic therapy (PDT), in which the two-photon photochemistry is used [22,26,32]. As explained in Section 3.3, in molecules with the two-photon photochemistry property, an electron is excited by a photon with a wavelength of λ_1 from S_0 to S_n state, then transfers to T_1 state, and is finally excited to T_n state by a photon with another wavelength of λ_2 (see Figure 3.8b). The electrons in the T_n state

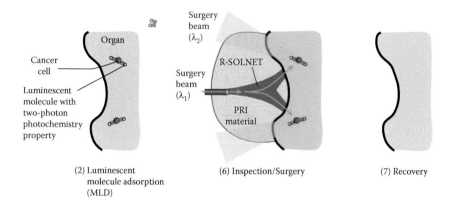

(2) Luminescent   (6) Inspection/Surgery   (7) Recovery
molecule adsorption
(MLD)

FIGURE 7.26   Concept of the SOLNET-assisted PDT utilizing the two-photon photochemistry.

induce chemical reactions for attacking cancer cells. This mechanism enables us to destroy cancer cells located in deep parts as mentioned below.

In the conventional PDT, it is hard for the excitation light to reach the deep parts. By using the two-photon photochemistry, three-dimensional attack on cancer cells might be possible because the chemical reactions occur only in regions, where, photons with λ_1 and photons with λ_2 coexist. Therefore, the chemical reactions can selectively be induced in any regions we want by controlling the positions of the two light beams with different wavelengths.

As illustrated in Figure 7.26, firstly, luminescent molecules with the two-photon photochemistry property are adsorbed by cancer cells. After forming R-SOLNETs that stretch toward the cancer cells, surgery beams of λ_1 are emitted from the R-SOLNETs toward the cancer cells. At the same time, surgery beams of λ_2 are introduced so that the λ_1 beams and the λ_2 beams overlap at the cancer cell sites. Thus, cancer cells located at deep parts are destroyed.

Although many molecules with the two-photon photochemistry property are known, such as porphyrin, biacetyl, comphorquinone, benzyl, etc. [33], in order to apply them to the human body, more advanced molecules that are solved in blood safely, have a selective adsorption capability, and exhibit light absorption for λ_1 and λ_2 in a range of 600–1000 nm, where light absorption due to hemoglobin is weak, should be searched. These methods are expected to be especially effective for selective removal of scattered small cancer cells without hurting the normal cells.

7.3.3   INDICATOR FOR REFLECTIVE OR LUMINESCENT MATERIALS USING R-SOLNET

Figure 7.27 shows an example of an indicator for reflective or luminescent materials using R-SOLNETs [26]. Between an optical waveguide and micro fluidic channels (MFCs), a PRI material is placed. From the optical waveguide, a write beam is introduced into the PRI material. If Material 1 and Material 3 are reflective (or luminescent), the write beam and the reflected write beam (or luminescence) are superposed to increase the refractive index in the beam overlapping regions. This effect induces self-focusing

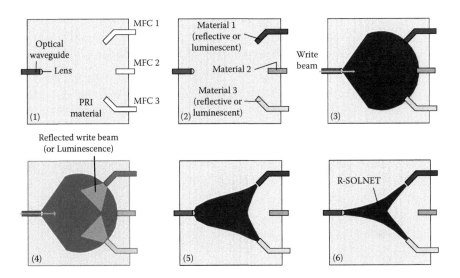

FIGURE 7.27 Indicator for reflective or luminescent materials using R-SOLNETs.

to make R-SOLNETs, namely, self-organized lightwave paths between the optical waveguide and MFC 1 and MFC 3. This enables us to find the locations of the reflective (or luminescent) materials as well as to transmit the write beam, the reflected write beam (or luminescence), and other light beams through the R-SOLNETs between the optical waveguide and the MFCs containing reflective or luminescent materials.

REFERENCES

1. T. Mikawa, M. Kinoshita, K. Hiruma, T. Ishitsuka, M. Okabe, S. Hiramatsu, H. Furuyama et al. "Implementation of Active Interposer for High-Speed and Low-Cost Chip Level Optical Interconnects," *IEEE J. Select. Topics in Quantum Electron.* **9**, 452–459, 2003.
2. B. S. Rho, S. Kang, H. S. Cho, H.-H. Park, S.-W. Ha, and B.-H. Rhee, "PCB-Compatible Optical Interconnection Using 45o-Ended Connection Rods and Via-Holed Waveguides," *J. Lightwave Technol.* **22**, 2128–2134, 2004.
3. A. Glebov, M. G. Lee, D. Kudzuma, J. Roman, M. Peters, L. Huang, and S. Zhou, "Integrated Waveguide Microoptic Elements for 3D Routing in Board-Level Optical Interconnects," *Proc. SPIE* **6126**, 61260L, 2006.
4. F. E. Doany, C. L. Schow, B. G. Lee, R. Budd, C. Baks, R. Dangel, R. John et al. "Terabit/sec-Class Board-Level Optical Interconnects Through Polymer Waveguides Using 24-Channel Bidirectional Transceiver Modules," *Proc. 61st Electronic Components & Technology Conference (ECTC)*, Lake Buena Vista, Florida, 790–797, 2011.
5. T. Yoshimura, J. Roman, Y. Takahashi, M. Lee, B. Chou, S.I. Beilin, W.V. Wang, and M. Inao, "Proposal of Optoelectronic Substrate with Film/Z-Connection Based on OE-Film," *Proc. 3rd IEMT/IMC Symposium*, Japan, 140–145, 1999.
6. T. Yoshimura, J. Roman, Y. Takahashi, W. V. Wang, M. Inao, T. Ishitsuka, K. Tsukamoto, K. Motoyoshi, and W. Sotoyama, "Self-Organizing Waveguide Coupling Method "SOLNET" and its Application to Film Optical Circuit Substrates," *Proc. 50th Electronic Components & Technology Conference (ECTC)*, Las Vegas, Nevada, 962–969, 2000.

7. T. Yoshimura, Y. Takahashi, M. Inao, M. Lee, W. Chou, S. Beilin, W.-C. Wang, J. Roman, and T. Massingill, "Systems Based on Opto-Electronic Substrates with Electrical and Optical Interconnections and Methods for Making," U.S. Patent 6,343,171 B1, 2002.

8. T. Yoshimura, M. Ojima, Y. Arai, and K. Asama, "Three-Dimensional Self-Organized Micro Optoelectronic Systems for Board-Level Reconfigurable Optical Interconnects— Performance Modeling and Simulation," *IEEE J. Select. Topics in Quantum Electron.* **9**, 492–511, 2003.

9. T. Yoshimura, T. Inoguchi, T. Yamamoto, S. Moriya, Y. Teramoto, Y. Arai, T. Namiki, and K. Asama, "Self-Organized Lightwave Network Based on Waveguide Films for Three-Dimensional Optical Wiring Within Boxes," *J. Lightwave Technol.* **22**, 2091– 2100, 2004.

10. T. Yoshimura, M. Miyazaki, Y. Miyamoto, N. Shimoda, A. Hori, and K. Asama, "Three-Dimensional Optical Circuits Consisting of Waveguide Films and Optical Z-Connections," *J. Lightwave Technol.* **24**, 4345–4352, 2006.

11. T. Yoshimura, *Optical Electronics: Self-Organized Integration and Applications*, Pan Stanford Publishing Pte. Ltd., Singapore, 2012.

12. B. Chou, S. Beilin, H. Jiang, D. Kudzuma, M. Lee, M. McCormack, T. Massingill et al. "Multilayer High Density Flex Technology," *Proc. 49th Electronic Components & Technology Conference (ECTC)*, San Diego, California, 1181–1189, 1999.

13. T. Yoshimura, K. Kumai, T. Mikawa, O. Ibaragi, and O. Bonkohara, "Photolithographic Packaging with Selectively Occupied Repeated Transfer (PL-Pack with SORT) for Scalable Optical Link Multi-chip-module (S-FOLM)," *IEEE Trans. Electron. Packag. Manufact.* **25**, 19–25, 2002.

14. T. Yoshimura, D. Tamaki, S. Kawamura, Y. Arai, and K. Asama, "A Device Transfer Process "Selectively Occupied Repeated Transfer (SORT)" for Resource-Saving Integration of Polymer Micro Optics Fabricated by "Built-in Mask" Method," *IEEE Trans. Comp., Packag. Technol.* **27**, 468–471, 2004.

15. T. Yoshimura, T. Kofudo, T. Kashiwazako, K. Naito, K. Ogushi, and Y. Kitabayashi, "A Material-Saving Optical Waveguide Fabrication Process with Selective Transfers of Cores," *Opt. Eng.* **47**, 014601, 2008.

16. T. Yoshimura, J. Roman, Y. Takahashi, S. I. Beilin, W. V. Wang, and M. Inao, "Optoelectronic Amplifier/Driver-Less Substrate for Polymer-Waveguide-Based Board-Level Interconnection -Calculation of Delay and Power Dissipation," *Nonlin. Opt.* **22**, 453–456, 1999.

17. T. Yoshimura, S. Tsukada, S. Kawakami, M. Ninomiya, Y. Arai, H. Kurokawa, and K. Asama, "Three-Dimensional Micro-Optical Switching System Architecture Using Slab-Waveguide-Based Micro-Optical Switches," *Opt. Eng.* **42**, 439–446, 2003.

18. T. Yoshimura, Y. Arai, H. Kurokawa, and K. Asama, "Predicted Insertion Loss Reductions Achieved by Implementing Three-Dimensional Micro Optical Network in Chip-Scale Optical Interconnects," *IEEE Photon. Technol. Lett.* **16**, 647–649, 2004.

19. N. Fujimoto, H. Tomita, and T. Yoshimura, "A Path-Independent-Loss Type Optical Cross-Connect System using High-Speed Waveguide Prism Deflector Micro-Optical Switches in WDM Metworks," *Research Reports of the School of Engineering, Kinki University* **40**, 81–85, 2006.

20. K. Takahashi, "Research and Development on Ultra-High-Density 3-Dimensional LSI-Chip-Stack Packaging Technologies," *The 3rd Annual Meeting on Electronics System Integration Technologies Digest* (edited and published by the Electronic System Integration Technology Research Department, Association of Super-Advanced Electronics Technologies (ASET)), Japan 43–94, 2002.

21. B. O'Regan and M. Gratzel, "A Low-Cost, High-Efficiency Solar Cell Based on Dye-Sensitized Colloidal TiO_2 Films," *Nature* **353**, 737–740, 1991.

22. T. Yoshimura, H. Watanabe, and C. Yoshino, "Liquid-Phase Molecular Layer Deposition (LP-MLD): Potential Applications to Multi-Dye Sensitization and Cancer Therapy," *J. Electrochem. Soc.* **158**, 51–55, 2011.

23. T. Yoshimura, R. Ebihara, and A. Oshima, "Polymer Wires with Quantum Dots Grown by Molecular Layer Deposition of Three Source Molecules for Sensitized Photovoltaics," *J. Vac. Sci. Technol. A* **29**, 051510, 2011.

24. T. Yoshimura and S. Ishii, "Effect of Quantum Dot Length on the Degree of Electron Localization in Polymer Wires Grown by Molecular Layer Deposition," *J. Vac. Sci. Technol. A* **31**, 031501, 2013.

25. T. Yoshimura, S. Bai, H. Tateno, and C. Yoshino, "*In situ* Photocurrent Spectra Measurements during Growth of Three-Dye-Stacked Structures by the Liquid-Phase Molecular Layer Deposition," *J. Appl. Phys.* **122**, 015309, 2017.

26. T. Yoshimura, *Thin-Film Organic Photonics: Molecular Layer Deposition and Applications*, CRC/Taylor & Francis, Boca Raton, Florida, 2011.

27. R. Shioya and T. Yoshimura, "Design of Solar Beam Collectors Consisting of Multi-Layer Optical Waveguide Films for Integrated Solar Energy Conversion Systems," *J. Renew. Sustain. Energy* **1**, 033106, 2009.

28. A. Jassim and T. Yoshimura, "Proposal of Integrated Sensitized Solar Cell Films Based on Sputtered ZnO Thin Films," *ECS Trans.* **69**, 193–197, 2015.

29. K. Sayama and H. Arakawa, "Dream of Artificial Synthesis," *Newton: Next-Generation Technology*, Newton Press, Tokyo, 102–109, September 2002 [in Japanese].

30. S. Sato, T. Arai, T. Morikawa, K. Uemura, TM. Suzuki, H. Tanaka, and T. Kajino, "Selective CO_2 Conversion to Formate Conjugated with H_2O Oxidation Utilizing Semiconductor/Complex Hybrid Photocatalysts," *J. Am. Chem. Soc.* **133**, 15240–15243, 2011.

31. T. Yoshimura, C. Yoshino, K. Sasaki, T. Sato, and M. Seki, "Cancer Therapy Utilizing Molecular Layer Deposition (MLD) and Self-Organized Lightwave Network (SOLNET) -Proposal and Theoretical Prediction," *IEEE J. Select. Topics in Quantum Electron.* **18**, 1192–1199, 2012.

32. T. Yoshimura, "In-Situ Drug Synthesis at Cancer Cells for Molecular Targeted Therapy by Molecular Layer Deposition -Conceptual Proposal," Chapter 17 *in Nano Based Drug Delivery*, IAPC Publishing, Croatia, 429–458, 2015.

33. C. Brauchle, U.P. Wild, D.M. Burland, G.C. Bjorkund, and D.C. Alvares, "Two-Photon Holographic Recording with Continuous-Wave Lasers in the 750-1100-nm Range," *Opt. Lett.* **7**, 177–179, 1982.

8 Future Challenges

Results described in the previous chapters revealed that the SOLNET is a promising candidate to realize the following functions.

1. Optical solder: Self-aligned optical couplings between misaligned optical devices with different core sizes
2. 3D optical wiring in free spaces
3. Targeted lightwave paths onto specific objects

Meanwhile, because the SOLNET is a technology in progress, a lot of issues to be solved are left. In this chapter, the future challenges to improve the SOLNET performance are discussed.

8.1 UNMONITORED SOLNET FORMATION

For practical use of the SOLNET optical solder, it is desirable that SOLNETs can be formed with fixed writing time regardless of the coupling conditions. From this viewpoint, one concern in SOLNETs is the misalignment dependence of the optimum writing time required to reach the maximum coupling efficiency. In Figure 8.1a, the current situation is schematically illustrated. With increasing the misalignments between coupled optical devices, the optimum writing time increases as mentioned in Sections 4.1.1 and 4.2.3. In such cases, we should have to monitor the coupling efficiency and adjust the writing time to obtain the maximum efficiency.

As described in Section 4.5.2, the optimum writing time difference can be suppressed by increasing the gap distance between optical devices. So, optimization of the gap distance might be a possible approach to obtain the Ideal Situation I shown in Figure 8.1b, where the optimum writing time is kept constant regardless of the misalignments. This enables unmonitored SOLNET formation. Ideal Situation II, where coupling efficiency versus writing time curves exhibit saturated characteristics, is another approach toward the unmonitored SOLNET formation.

Optimization of gamma characteristics of the PRI materials might contribute to reducing the optimum writing time difference and realizing the saturated coupling efficiency.

In two-photon SOLNETs, as mentioned in Section 4.2.3, there is another concern, that is, the rapid decrease in the coupling efficiency with writing time observed after a maximum has been reached, which narrows the writing time windows in coupling efficiency. The rapid decrease might be caused by the zigzag SOLNET structures that appear after long exposure to the write beams as can be seen in Figure 4.18b. The zigzag structures are believed to be a consequence of the steep formation characteristics of the two-photon SOLNET (see Figure 4.19).

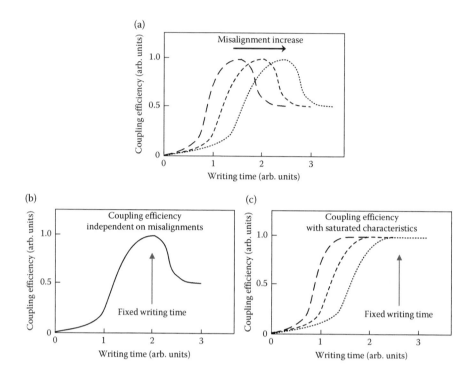

FIGURE 8.1 Dependence of the coupling efficiency vs. writing time characteristics on misalignments. (a) Current situation, (b) Ideal situation I, and (c) Ideal situation II.

One of the approaches to widen the writing time windows might be a slowdown in the SOLNET formation speed by reducing the write-beam intensity or the PRI material sensitivity (see Figure 4.49). From a viewpoint of materials, optimization of gamma characteristics of the PRI materials is expected to contribute to widening the writing time window.

8.2   CONTROL OF GAMMA CHARACTERISTICS OF PRI MATERIALS

Figure 8.2 shows an approach to control the gamma characteristics of PRI materials. The material consists of three component molecules; a molecule with high refractive index (high-n molecule), a molecule with middle refractive index (middle-n molecule), and a molecule with low refractive index (low-n molecule). In the case that the photochemical sensitivity, γ, of the middle-n molecule is larger than that of the high-n molecule, with write-beam exposure, the refractive index initially increases with a low rate, and then increases with a high rate. In the case that γ of the middle-n molecule is smaller than that of the high-n molecule, the refractive index increases with a high rate, and then increases with a low rate. By using more than three component molecules, more precise control of the gamma characteristics might be achieved.

The gamma characteristics control for the PRI materials is expected to suppress the SOLNET degradation caused by the overexposure such as the broadening and

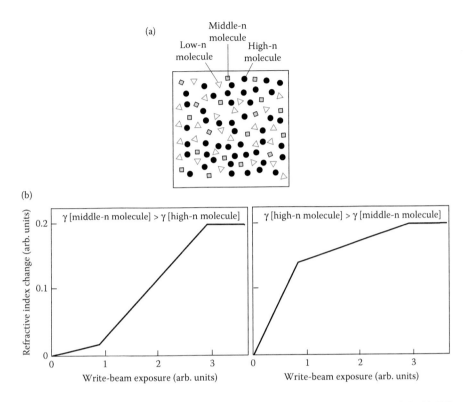

FIGURE 8.2 Approach to the control of the gamma characteristics of PRI materials. (a) PRI material consisting of three component molecules and (b) expected gamma characteristics.

the zigzag appearance, and contribute to widening the writing time windows. By precise control of the gamma characteristics, reduction of the optimum writing time difference shown in Figure 8.1b, and the coupling efficiency with the saturated characteristics shown in Figure 8.1c might be achieved.

Simulations for the effect of the gamma characteristics on SOLNET formation are interesting future work.

8.3 DYNAMIC SOLNETs

When usual photosensitive materials such as photopolymers, photodefinable materials, and photosensitive glass are used for the PRI materials, SOLNETs provide permanent lightwave paths. In order to produce dynamic SOLNETs, which enable reorganization of optical wiring as schematically illustrated in Figure 8.3, rewritable PRI materials, for example, photorefractive materials and third-order nonlinear optical materials shown in Figure 3.7b and c are required. Fazio et al. have already developed the soliton waveguides in the photorefractive crystals of lithium niobate, opening the way to the dynamic optical interconnects as mentioned in Section 2.2. Development of high-speed rewritable PRI materials driven by low-power light beams might be the future challenge.

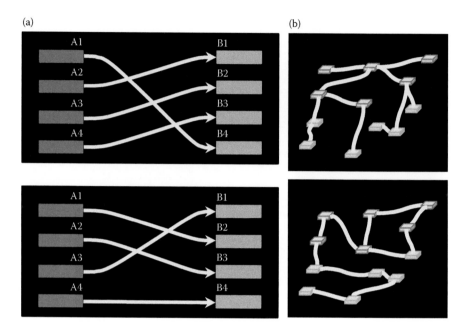

FIGURE 8.3   Concept of (a) the 2D dynamic SOLNETs and (b) the 3D dynamic SOLNETs.

8.4   OTHERS

Finally, other challenges are listed as follows.

- Investigation of angular misalignments: In the die-bonding process, as well as the lateral misalignment, the angular misalignment is an issue. Systematic FDTD-based simulations on the effect of the angular misalignments on SOLNET formation should be performed.
- Simulations based on the 3D FDTD Method: Because SOLNET formation is basically a 3D process, simulations based on the 3D FDTD method should be the future work.
- Experimental demonstrations of SOLNETs in nanoscale optical circuits.
- Optimization of wavefront shapes and spread angles of write beams to improve the SOLNET performance. For example, structural optimization of luminescent targets and waveguide core edges, lens insertion, and so on.
- Developments of High-Δn PRI materials for high-index-contrast SOLNETs.
- Investigation of SOLNETs accompanying interference patterns.

Appendix I: Methods of Computer Simulations for SOLNETs

AI.1 ANALOGY BETWEEN PHOTONS AND ELECTRONS

To understand the behaviors of lightwaves, it is convenient to consider the analogy between photons and electrons. It is known that the electron wavefunction ψ and the electric field of a lightwave $\mathbf{E}$ are determined by the following wave equations, respectively.

Time-Independent Schrödinger's Equation:

$$\left(-\frac{\hbar^2}{2m}\nabla^2 + V(\mathbf{x})\right)\psi(\mathbf{x}) = E\psi(\mathbf{x}) \tag{AI.1}$$

Helmholtz Equation:

$$\left(\nabla^2 + (n(\mathbf{x})k_0)^2\right)\mathbf{E}(\mathbf{x}) = 0 \tag{AI.2}$$

Here, m, V, and E are mass, potential energy, and total energy of an electron. $\hbar$ is the Plank constant divided by 2π. n is the refractive index, and k_0 is the wavenumber of a lightwave in a vacuum. Equations AI.1 and AI.2 can be rewritten as Equation AI.3.

$$\left.\begin{aligned}\left(-\frac{\hbar^2}{2m}\nabla^2 - (E - V(\mathbf{x}))\right)\psi(\mathbf{x}) = 0 \\ \left(-\frac{\hbar^2}{2m}\nabla^2 - \frac{\hbar^2 k_0^2}{2m}n^2(\mathbf{x})\right)\mathbf{E}(\mathbf{x}) = 0\end{aligned}\right\} \tag{AI.3}$$

By comparing these equations, the following correspondence is found.

$$\psi(\mathbf{x}) \leftrightarrow \mathbf{E}(\mathbf{x}) \tag{AI.4}$$

$$E - V(\mathbf{x}) \leftrightarrow \frac{\hbar^2 k_0^2}{2m}n^2(\mathbf{x}) \tag{AI.5}$$

ψ represents the probability amplitude for an electron, namely, $\psi^*\psi$ is proportional to the probability to find the electron. The relationship of Equation AI.4 leads us to believe that **E** represents the probability amplitude for a photon, namely, **E***E** is proportional to the probability to find the photon.

As shown in Figure AI.1a, an electron tends to be confined in a region with small potential energy, for example, in a quantum well. The relationship of Equation AI.5 indicates that "$-V(\mathbf{x})$" corresponds to "$n^2(\mathbf{x})$." This suggests that a photon tends to be confined in a region with a high refractive index, for example, in a core. The core and cladding for a photon respectively correspond to the quantum well and barrier for an electron. Therefore, by constructing a line-shaped region with a higher refractive index in a planar substrate, a photon is confined in the region as shown in Figure AI.1b. The line-shaped region is the core of an optical waveguide, and the surrounding area is the cladding. By forming the higher-refractive-index regions with designated patterns in the substrate, photons can be guided through the designated routes to make optical circuits.

For analyses of optical circuits, the finite difference time domain (FDTD) method and the beam propagation method (BPM) are useful. The FDTD method calculates electric fields **E** and magnetic fields **H** by solving the Maxwell's equations. The method can treat any light beams including reflected/emitted light beams and widely-spread light beams, although it requires a large memory capacity and long calculation time, limiting the model size.

The BPM calculates light-beam propagation along a propagation axis. The method is suitable for large size models since it can perform high-speed calculations with a small memory capacity. However, the BPM cannot treat off-axis light-beam propagation such as reflected/emitted light beams and widely-spread light beams.

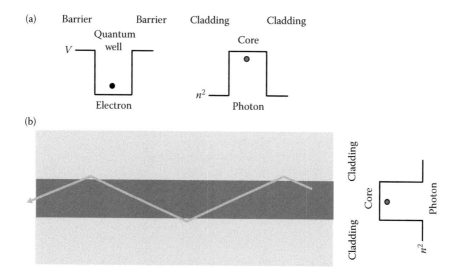

FIGURE AI.1 (a) Analogy between the electron confinement and the photon confinement, and (b) schematic illustration of the photon confinement in an optical waveguide.

Therefore, it is necessary for us to use these two methods appropriately depending on the models.

In the present Appendix, the FDTD method and the BPM using the Fourier transform are briefly reviewed.

AI.2   FINITE DIFFERENCE TIME DOMAIN (FDTD) METHOD

In FDTD method, electric fields **E** and magnetic fields **H** are calculated by solving the Maxwell's equations directly. The starting equations are:

$$\frac{\partial \mathbf{E}}{\partial t} = -\frac{\sigma}{\varepsilon}\mathbf{E} + \frac{1}{\varepsilon}\nabla \times \mathbf{H} \tag{AI.6}$$

$$\frac{\partial \mathbf{H}}{\partial t} = -\frac{1}{\mu}\nabla \times \mathbf{E}. \tag{AI.7}$$

Here, σ, ε, and μ are conductivity, dielectric constant, and magnetic permeability, respectively. Difference equations for the differential equations, Equations AI.6 and AI.7, can be written as follows.

$$\mathbf{E}^n = \frac{\left(1 - \dfrac{\sigma \Delta t}{2\varepsilon}\right)}{\left(1 + \dfrac{\sigma \Delta t}{2\varepsilon}\right)}\mathbf{E}^{n-1} + \frac{\dfrac{\Delta t}{\varepsilon}}{\left(1 + \dfrac{\sigma \Delta t}{2\varepsilon}\right)}\nabla \times \mathbf{H}^{n-\frac{1}{2}} \tag{AI.8}$$

$$\mathbf{H}^{n+\frac{1}{2}} = \mathbf{H}^{n-\frac{1}{2}} - \frac{\Delta t}{\mu}\nabla \times \mathbf{E}^n \tag{AI.9}$$

n represents time steps, meaning that $t = n\Delta t$ when the time interval for a step is Δt.

Using Equation AI.8, electric fields at t are determined by information at a previous time, namely, electric fields at $t - \Delta t$ and magnetic fields at $t - (1/2)\Delta t$. Similarly, using Equation AI.9, magnetic fields at $t + (1/2)\Delta t$ are determined by magnetic fields at $t - (1/2)\Delta t$ and electric fields at t. The calculation sequences in the FDTD method is summarized in Figure AI.2.

In actual calculations, we need to modify Equations AI.8 and AI.9 to give difference equations for positions in addition to time. For two-dimensional models in the x-y plane with polarization//z, the difference equations for E_z^n, $H_x^{n+1/2}$, $H_y^{n+1/2}$ are given as follows.

$$E_z^n(i,j) = C_{EZ}(i,j)E_z^{n-1} + C_{EZIX}(i,j)\left(H_y^{n-\frac{1}{2}}\left(i+\frac{1}{2},j\right) - H_y^{n-\frac{1}{2}}\left(i-\frac{1}{2},j\right)\right)$$
$$- C_{EZIY}(i,j)\left(H_x^{n-\frac{1}{2}}\left(i,j+\frac{1}{2}\right) - H_x^{n-\frac{1}{2}}\left(i,j-\frac{1}{2}\right)\right) \tag{AI.10}$$

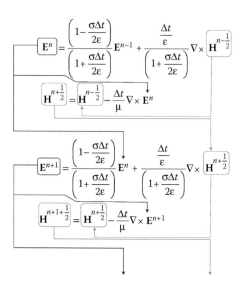

FIGURE AI.2   Calculation sequences in the FDTD method.

$$C_{EZ}(i,j) = \frac{\left(1 - \dfrac{\sigma(i,j)\Delta t}{2\varepsilon(i,j)}\right)}{\left(1 + \dfrac{\sigma(i,j)\Delta t}{2\varepsilon(i,j)}\right)} \qquad C_{EZIX}(i,j) = \frac{\dfrac{\Delta t}{\varepsilon(i,j)}}{\left(1 + \dfrac{\sigma(i,j)\Delta t}{2\varepsilon(i,j)}\right)}\frac{1}{\Delta x} \qquad (AI.11)$$

$$C_{EZIY}(i,j) = \frac{\dfrac{\Delta t}{\varepsilon(i,j)}}{\left(1 + \dfrac{\sigma(i,j)\Delta t}{2\varepsilon(i,j)}\right)}\frac{1}{\Delta y}$$

$$H_x^{n+\frac{1}{2}}\left(i,j+\frac{1}{2}\right) = H_x^{n-\frac{1}{2}}\left(i,j+\frac{1}{2}\right) - C_{HXIY}\left(i,j+\frac{1}{2}\right)\left(E_z^n(i,j+1) - E_z^n(i,j)\right) \quad (AI.12)$$

$$C_{HXIY}\left(i,j+\frac{1}{2}\right) = \frac{\Delta t}{\mu\left(i,j+\dfrac{1}{2}\right)}\frac{1}{\Delta y} \qquad (AI.13)$$

$$H_y^{n+\frac{1}{2}}\left(i+\frac{1}{2},j\right) = H_y^{n-\frac{1}{2}}\left(i+\frac{1}{2},j\right) + C_{HYIX}\left(i+\frac{1}{2},j\right)\left(E_z^n(i+1,j) - E_z^n(i,j)\right) \quad (AI.14)$$

$$C_{HYIX}\left(i+\frac{1}{2},j\right)=\frac{\Delta t}{\mu\left(i+\frac{1}{2},j\right)}\frac{1}{\Delta x} \tag{AI.15}$$

Here, i and j represent positional steps along the x-axis and y-axis, respectively, meaning that $x = i\Delta x$ and $y = j\Delta y$.

As shown in Figure AI.3a, $E_z(i,j)$ at time step n can be determined by surrounding $H_x(i,j-1/2)$, $H_x(i,j+1/2)$, $H_y(i-1/2,j)$, and $H_y(i+1/2,j)$ as well as $E_z(i,j)$ at previous time steps. For $H_x(i,j+1/2)$ at time step $n+1/2$, as shown in Figure AI.3b, it is determined by $E_z(i,j)$ and $E_z(i,j+1)$ as well as $H_x(i,j+1/2)$ at previous time steps. For $H_y(i+1/2j)$ at time step $n+1/2$, as shown in Figure AI.3c, it is determined by $E_z(i,j)$ and $E_z(i+1,j)$ as well as $H_y(i+1/2,j)$ at previous time steps. By repeating the procedure, electric field and magnetic field distributions all over the calculation region can be obtained as a function of time.

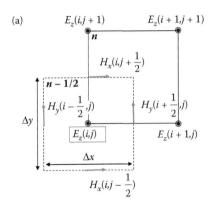

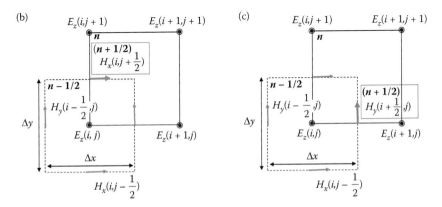

FIGURE AI.3 Diagrams for calculations of (a) E_z, (b) H_x, and (c) H_y.

AI.3   BEAM PROPAGATION METHOD (BPM)

In Figure AI.4, a schematic illustration for simulations of light-beam propagation by the BPM is shown. Waveguide cores have a refractive index (n_1) higher than that of the surrounding cladding region (n_2). Electric fields of light beams introduced into the calculation region are successively calculated along the propagation axis (z-axis) with a step of Δz to simulate the light-beam propagation following the procedure mentioned below.

Figure AI.5a shows a model for calculations by the BPM using the Fourier transform [1]. The whole calculation region with a width of W is regarded as a virtual waveguide with a width of W. Virtual modes in the virtual waveguide schematically shown in Figure AI.5b are expressed as follows.

$$E_\nu(x,z) = F_\nu \sin K_\nu x \cdot e^{-in_2 k_0 z}, \quad F_\nu = \sqrt{\frac{4\mu_0\omega}{n_2 k_0 W}}, \quad K_\nu = \frac{\pi\nu}{W}, \quad \nu = 1,2,3,\cdots \quad (AI.16)$$

Here, ν represents the mode number and ω the angular frequency of the lightwave. μ_0 and k_0 are respectively magnetic permeability and wavenumber in

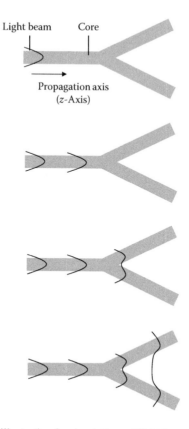

FIGURE AI.4   Schematic illustration for simulations of light-beam propagation by the BPM.

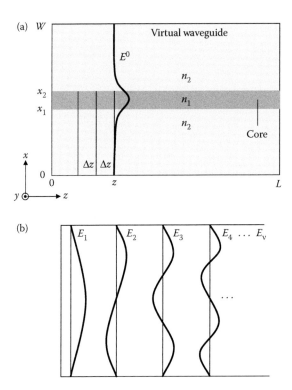

FIGURE AI.5   (a) A model for calculations by the BPM, and (b) virtual modes in a virtual waveguide.

a vacuum. The electric field distributions in the calculation region, denoted by E^0, can be expressed by the superposition of the electric field distributions of the virtual modes as follows.

$$E^0(x,z) = \sum_{\nu=1}^{\infty} A_\nu(z)E_\nu(x,z) \qquad (AI.17)$$

$A_\nu(z)$ is the weight coefficient for each $E_\nu(x,z)$. $A_\nu(z)$ and its derivative can be written as:

$$A_\nu(z) = \frac{\displaystyle\int_0^W E_\nu^*(x,z)E^0(x,z)dx}{\displaystyle\int_0^W E_\nu^*(x,z)E_\nu(x,z)dx}, \qquad (AI.18)$$

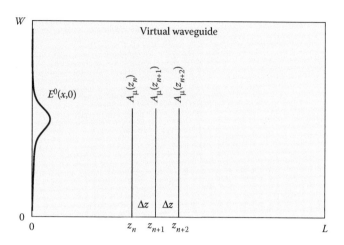

FIGURE AI.6 Calculation procedure in the BPM.

$$\frac{dA_\mu(z)}{dz} = \frac{n_2^2 k_0^2 \sum_{\nu=1}^{M} A_\nu(z) \int_0^W f(x,z) E_\mu^*(x,z) E_\nu(x,z) dx - K_\mu^2 A_\mu(z) \int_0^W E_\mu^*(x,z) E_\mu(x,z) dx}{i2n_2 k_0 \int_0^W E_\mu^*(x,z) E_\mu(x,z) dx}.$$

(AI.19)

$f(x,z)$ is given by,

$$f(x,z) = \frac{n^2(x,z)}{n_2^2} - 1,$$

(AI.20)

where, $n(x,z)$ is the refractive index in the calculation region.

As Figure AI.6 shows, the weight coefficient at z_{n+2} is determined by information at previous steps, namely, the weight coefficient at z_n and slope of the weight coefficient along z-axis at z_{n+1}, by the following equation.

$$A_\mu(z_{n+2}) = A_\mu(z_n) + \frac{dA_\mu(z)}{dz}\bigg|_{z=z_{n+1}} (2\Delta z)$$

(AI.21)

By repeating the procedure from $z = 0$ to L, which is the end point of the calculation region, $A_\nu(z)$ at $z = 0$ to L is calculated. Finally, by substituting $A_\nu(z)$ into Equation AI.17, electric field distributions all over the calculation region can be obtained.

REFERENCE

1. M. Seino, H. Nakajima, Y. Daido, I. Sawaki, and K. Asama, "Optical Waveguide Analysis Using the Fourie Transform and Its Application to Intersecting Waveguides," *IEICE Ronbunshi* **J-66C**, 732–739, 1985 [in Japanese].

Appendix II: Molecular Layer Deposition (MLD)

Molecular layer deposition (MLD) is a monomolecular-step growth process for organic thin films with designated molecular arrangements [1–4]. MLD utilizes selective chemical reactions (or electrostatic force) between different kinds of molecules. As Figure AII.1 shows, firstly, Molecule A is provided onto a substrate surface to form a monomolecular layer of Molecule A. By switching molecules from A to B, Molecule B is connected to Molecule A. Once the surface is covered with Molecule B, the film growth is automatically terminated by the self-limiting effect similar to that in the Atomic layer deposition (ALD) [5]. By switching molecules in a sequence of A, B, C, D, and so on, an organic thin film with designated molecular arrangements of A/B/C/D is obtained. MLD can be performed both in vapor phase [1–4] and in liquid phase [3,6].

Figure AII.2 shows an experimental demonstration of MLD using pyromellitic dianhydride (PMDA) and 4,4'-diaminodiphenyl ether (DDE) [1]. When PMDA is provided onto a DDE surface, the film thickness, which is monitored by a quartz crystal microbalance, rapidly increases and then is saturated in 120 s. When the molecules are switched from PMDA to DDE, again, the film thickness rapidly increases and is saturated. By repeating the molecule switching, step-like film growth is observed. The thickness change for one growth step is close to the sizes of PMDA and DDE. These results indicate that monomolecular-step growth is performed by MLD.

Three featured capabilities of MLD are schematically depicted in Figure AII.3.

1. *Ultra-thin/conformal organic material growth*: MLD enables ultra-thin/ conformal organic material growth on three-dimensional surfaces with arbitrary structures including surfaces with ultra-fine patterns, particles, and porous objects.
2. *Tailored organic material growth*: MLD enables tailored organic material growth with artificially-controlled molecular sequences.
3. *Selective organic material growth*: Utilizing patterned surface treatments, MLD enables selective organic material growth.

The capability of the ultra-thin/conformal organic material growth has produced many applications. Zhou and Bent succeeded to form copper diffusion barriers in LSIs and to deposit photoresist by MLD [4]. Liang and Weimer realized uniform polymer film coating on surfaces of nanoparticles [7]. George, Yoon, and Dameron developed a growth process for hybrid organic-inorganic polymer films [8] by combining MLD with ALD and the films were used as gate insulators for organic thin-film transistors [9], thin-film encapsulations for organic light-emitting diodes [10], and energy technologies including batteries and thermoelectric energy devices [11].

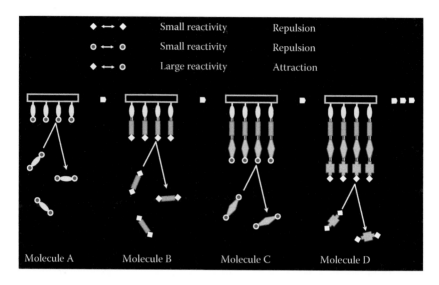

FIGURE AII.1   Concept of MLD.

The capability of the tailored organic material growth realizes artificial controls of the electron wavefunction shapes in organic thin films to optimize the material performance or to generate new functions [3,12,13]. For example, as schematically shown in Figure AII.4, widely spread wavefunctions, symmetric wavefunctions, and non-centrosymmetric wavefunctions can be formed in polymer wires or molecular stacked structures by controlling the molecular sequences.

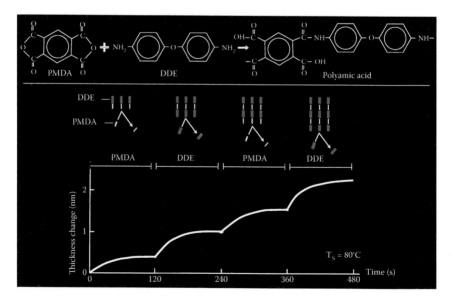

FIGURE AII.2   Experimental demonstration of MLD.

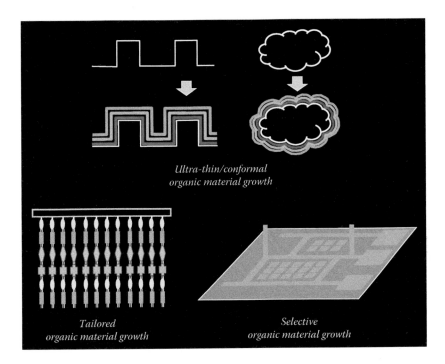

FIGURE AII.3 Featured capabilities of MLD.

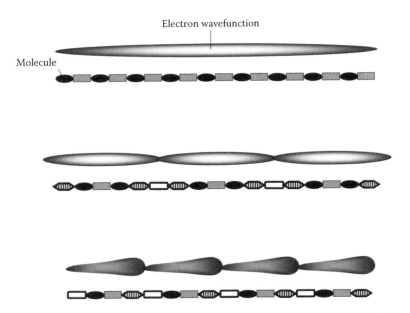

FIGURE AII.4 Control of electron wavefunction shapes by placing constituent molecules in designated arrangements.

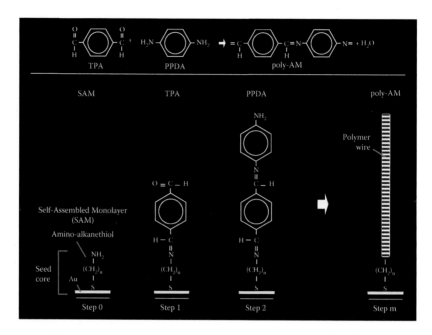

FIGURE AII.5 MLD process for the location/orientation-controlled growth using seed cores for selective growth.

The capability of the selective organic material growth is available using surface treatments [14–16]. Figure AII.5 shows the process for the location/orientation-controlled growth using seed cores [15]. First, a self-assembled monolayer (SAM), such as amino-alkanethiol with amino groups on the top, is formed on a patterned Au film to fabricate a seed core. Next, terephthalaldehyde (TPA) is provided on the surface to connect it on the SAM. Then, p-phenylenediamine (PPDA) is provided to connect it to TPA. By repeating such growth steps, polymer wires of poly-azomethine (AM) can be grown nearly vertically on the seed core surface.

It is expected that by distributing seed cores appropriately, as Figure AII.6 shows, three-dimensional polymer wire networks would be constructed [3]. This

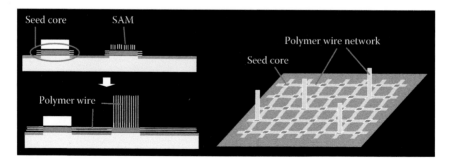

FIGURE AII.6 Expected structures of polymer wire network grown by the location/orientation-controlled growth.

approach would be promising to apply MLD to integrated photonic/electronic devices.

Selective growth can also be achieved by the hydrophilic/hydrophobic patterning [14,16].

REFERENCES

1. T. Yoshimura, S. Tatsuura, and W. Sotoyama, "Polymer Films Formed with Monolayer Growth Steps by Molecular Layer Deposition," *Appl. Phys. Lett.* **59**, 482–484, 1991.
2. T. Yoshimura, E. Yano, S. Tatsuura, and W. Sotoyama, "Organic Functional Optical Thin Film, Fabrication and Use Thereof," US Patent 5,444,811, 1995.
3. T. Yoshimura, *Thin-Film Organic Photonics: Molecular Layer Deposition and Applications*, CRC/Taylor & Francis, Boca Raton, Florida, 2011.
4. H. Zhou and S. F. Bent, "Fabrication of Organic Interfacial Layers by Molecular Layer Deposition: Present Status and Future Opportunities," *J. Vac. Sci. Technol. A* **31**, 040801, 2013.
5. T. Suntola, "Atomic Layer Epitaxy," *Material Science Reports* **4**(7), Elsevier Science Publishers, 1989.
6. T. Yoshimura, S. Bai, H. Tateno, and C. Yoshino, "*In Situ* Photocurrent Spectra Measurements During Growth of Three-Dye-Stacked Structures by the Liquid-Phase Molecular Layer Deposition," *J. Appl. Phys.* **122**, 015309, 2017.
7. X. Liang and A. Weimer, "Photoactivity Passivation of TiO_2 Nanoparticles Using Molecular Layer Deposited (MLD) Polymer Films," *J. Nanopart. Res.* **12**, 135–142, 2010.
8. S. M. George, B. Yoon, and A. A. Dameron, "Surface Chemistry for Molecular Layer Deposition of Organic and Hybrid Organic-Inorganic Polymers," *Acc. Chem. Res.* **42**, 498–508, 2009.
9. B. H. Lee, K. H. Lee, S. Im, and M. M. Sung, "Molecular Layer Deposition of ZrO_2-Based Organic–Inorganic Nanohybrid Thin Films for Organic Thin Film Transistors," *Thin Solid Films* **517**, 4056–4060, 2009.
10. D. Yu, Y. Yang, Z. Chen, Y. Tao, and Y. Liu, "Recent Progress on Thin-Film Encapsulation Technologies for Organic Electronic devices," *Optics Commun.* **362**, 43–49, 2016.
11. A. J. Karttunen, T. Tynell, and M. Karppinen, "Layer-by-Layer Design of Nanostructured Thermoelectrics: First-Principles Study of ZnO: Organic Superlattices Fabricated by ALD/MLD," *Nano Energy* **22**, 338–348, 2016.
12. T. Yoshimura, "Enhancing Second-Order Nonlinear Optical Properties by Controlling the Wave Function in One-Dimensional Conjugated Molecules," *Phys. Rev. B* **40**, 6292–6298, 1989.
13. T. Yoshimura, R. Ebihara, and A. Oshima, "Polymer Wires with Quantum Dots Grown by Molecular Layer Deposition of Three Source Molecules for Sensitized Photovoltaics," *J. Vac. Sci. Technol. A* **29**, 051510, 2011.
14. T. Yoshimura, N. Terasawa, H. Kazama, Y. Naito, Y. Suzuki, and K. Asama, "Selective Growth of Conjugated Polymer Thin Films by the Vapor Deposition Polymerization," *Thin Solid Films* **497**, 182–184, 2006.
15. T. Yoshimura, S. Ito, T. Nakayama, and K. Matsumoto, "Orientation-Controlled Molecule-by-Molecule Polymer Wire Growth by the Carrier-Gas-Type Organic Chemical Vapor Deposition and the Molecular Layer Deposition," *Appl. Phys. Lett.* **91**, 033103, 2007.
16. T. Yoshimura and S. Ishii, "Effect of Quantum Dot Length on the Degree of Electron Localization in Polymer Wires Grown by Molecular Layer Deposition," *J. Vac. Sci. Technol. A* **31**, 031501, 2013.

Epilogue

In January 2009, I had two offers to write books on different subjects. One of them was from CRC/Taylor & Francis for a book entitled *Thin-Film Organic Photonics: Molecular Layer Deposition and Applications*. I was afraid that it might be very hard to complete two books at the same time. So, I told my wife, Yoriko, "How do you think about writing two books at the same time in English?" Then, she said to me "I know that you have accomplished anything you once told me." Her answer made me decide to accept the two offers.

During the preparation of these books, Yoriko passed away on September 27, 2009. Yoriko continued to receive encouragements from her friends of JOSHIGAKUIN high school, Musashino Academia Musicae, Morimura-Gakuen parents, students of her piano school, and her relatives and family for five months before September 27.

From the left: Tetsuzo, Yoriko, Fumiko (Kawamura), Naoko, and Chikako
at Trellis Square Apartments, Sunnyvale, California, circa 1999.

Last year, CRC/Taylor & Francis kindly gave me another offer to write a book on SOLNETs. I believe that this fortunate chance is a result of Yoriko's help. I thank Yoriko for her sincere support and encouragement since 1977, and have placed her photographs with her family in the epilogue as a memorial to her.

Yoriko Yoshimura, a pianist, at SFO.

I should also note how my students gave great contributions to this book. Although the concept of the SOLNET was proposed during a period when I was at Fujitsu Laboratories, Ltd., only a limited number of results could be obtained there. After I moved to Tokyo University of Technology, some of the students in my laboratory strongly promoted the research to get a lot of interesting and important results. Therefore, most of the results described in this book are those that the students, whose names can be found in references presented in every chapter, got with their excellent efforts and passions. I am sincerely proud of them.

Yoriko and I hope that SOLNETs will contribute to some scientific fields in the future.

Tetsuzo Yoshimura
September 27, 2017

Index

#0002 - 280318 - C12 - 234/156/12 - PB - 9781138746886